The Political Economy of Coal

The Political Economy of Coal

Ferdinand E. Banks
The University of Uppsala, Sweden
The University of Melbourne, Australia

Lexington Books
D.C. Heath and Company/Lexington, Massachusetts/Toronto

Library of Congress Cataloging in Publication Data

Banks, Ferdinand E.
The political economy of coal.

Bibliography: p.
Includes index.
1. Coal trade. 2. Coal trade—Political aspects.
3. International economic relations. I. Title.
HD9540.5.B36 1985 338.2'724 82-48522
ISBN 0-669-06169-7 (alk. paper)

Published simultaneously in Canada
Printed in the United States of America on acid-free paper
International Standard Book Number: 0-669-06169-7
Library of Congress Catalog Card Number: 82-48522

For my parents, Lola T. and Ferdinand E. Banks, Sr., and my sister, Mrs. Ann Banks Tunnell

For my teachers at Illinois Institute of Technology, Roosevelt University, and the University of Stockholm, and also at Fort Jackson, South Carolina, and Ford Ord, California.

For my students at Stockholm, Uppsala, Dakar, Lisbon, and Sydney, and also at Fort Lewis; Camp Gifu, Japan; and Hardt and Bismarck Kasernes (Schwabish Gmuend)

and for my wife and children

Contents

Figures

Tables

Preface and Acknowledgments

This is a book on applied energy economics. Coal is the principal topic, but the reader will also find self-contained and thorough surveys of the world oil market, nuclear energy and uranium, and natural gas. A chapter on electricity is especially designed for teaching purposes. The final chapter contains a short summary of the book in the form of an up-to-date analysis of the world coal market. The style, though perhaps not the exact tone, of this study is derived from my book, *The Political Economy of Oil,* and as in that book I have tried here to keep complicated algebraic materials separate from the main text. The only exceptions are chapter 7 and one passage in chapter 4.

At this time I would like to name and thank some of the people and organizations that made this book possible. First on my list are the Tore Browaldh and Jan Wallander Foundations of Svenska Handelsbanken that financed my travel, and Dr. Duncan Ironmonger of the Institute for Applied Social and Economic Research of the University of Melbourne who arranged for my research fellowship at that university. I would also like to thank my many friends and colleagues at Melbourne University and the University of New South Wales, and especially Warrane College (Sydney), for their company, conversation, advice, and hospitality.

Thanks also go to Ken Walker of the Australian School of the Environment for organizing my lectures at Griffith University, Brisbane; to Peter Bohm and Marian Radetzki of Stockholm University, and Stephen Derrick of Melbourne University; and to the students in my energy economics courses at Uppsala University and the University of Stockholm. I would also like to express my gratitude to the students in my courses in international monetary economics at Uppsala University who have provided me with some invaluable mental gymnastics over the past six or seven years, and to the good people at Lexington Books who have suffered through an unusually long delay before receiving this book. Let me also point out that I have received important comments on some of the material in this book at the 1982 and 1984 meetings of the International Association of Energy Economists, Cambridge University, and from people who attended my seminar at the Resources Systems Institute of the East-West Center, Honolulu, Hawaii. Finally, I would like to thank the editors and publishers of *Energy Policy* and *The Chemical Economy and Engineering Review* (Tokyo) for permitting me to reproduce materials first published in those journals, and also the OPEC Quarterly Review.

1
Background: The Oil Decade

Even at the end of 1984 it is impossible for us to know whether October 1973 was a turning point in modern history, but it is indisputable that some fundamental changes in the scheme of things were initiated in that month. Aside from the panic that accompanied the first oil price escalation, perhaps the most vivid recollection that many of us have of this era was the general failure by economists and politicians to comprehend the cause and significance of the appearance of OPEC, or what could be expected from this and similar organizations in the future. In my own work, and particularly my book, *The Political Economy of Oil,* I took a generally skeptical view of this inexplicable shortcoming, and certainly I have no reason to change my tune at the present time. The oil price rise has at last been checked, at least for the time being, but at a cost of 33 million unemployed in the OECD countries, as well as a dramatic deceleration in the economic growth of the industrial world that is being felt in almost every corner of the Third World. Despite the recent business cycle upturn in the United States, which was fueled by consumption and not investment, the general economic outlook for the industrial world does not appear promising, and one thing has become crystal clear: If our political masters know how to regain and maintain the rates of growth of income and employment that existed prior to 1973, they are keeping this information to themselves.

The analysis that I believe is appropriate for explaining the working of the oil market turns on the near irrelevance of OPEC as a decisive factor in the 1973–1974 oil price run-up and the importance of market fundamentals. In addition, I like to depart company with those outraged persons who place the full blame for the first oil price shock on the major oil companies—although, like most intelligent people, I do have some curiosity as to why the possibility of a rapid price rise or a successful OPEC takeover of their producing assets occupied so small a role in the thinking of Big Oil's directors and advisors. Beginning in the mid-1960s world oil demand began growing rapidly, while non-OPEC energy production was stagnant in the wake of rapidly increasing extraction, investment, and distribution costs. The rate of growth of demand for OPEC oil was even larger than the corresponding figure for oil in general. In the 1965–1973 period oil consumption increased at an average annual rate of 7.7 percent in the noncentrally planned world, or about 2.4 percentage points faster than energy growth in general. Correspondingly, the demand for OPEC oil increased by almost 10 percent annually between 1969 and 1973. Had growth rates of this magnitude continued, the annual demand for oil in the noncentrally planned world would be almost 100 million barrels at the present time. As I have shown in *The Political Economy of Oil,* obtaining that much crude would have entailed enormous price rises, even without the presence of OPEC; and it is comparatively easy to develop both theoretical and empirical support for the position that the major purpose of these price rises would have been to suppress demand, rather than to have promoted more exploration and production.

If we can assume that the world economy might, at some point during this decade, reattain a semblance of normality, then a question should be raised as to the future of the demand for oil. Immediately after the first oil price shock the demand for oil and other energy materials fell dramatically; but by 1978 demand was back at the 1973 level, and in 1979 the total world demand for crude oil was higher than at any point in human history, or about 63 million barrels per day (63 Mbbl/d). Of this amount 51.5 Mbbl/d was demanded by the noncentrally planned countries, as compared to 44 Mbbl/d in 1983, while OPEC produced a total of 31.5 Mbbl/d of crude oil and natural gas liquids, as compared to a present output of about 18 Mbbl/d. It was during this period that Sheik Yamani, as well as some scenarios of the Workshop on Alternative Energy Strategies (WAES), predicted that the demand for OPEC oil would reach 45 Mbbl/d within ten years, and the president of at least one very large independent oil company claimed that by the turn of the century, oil would be selling for $100 a barrel. Although Professor Milton Friedman seems to think otherwise (having stated in *Newsweek,* March 21, 1983, that an oil glut was imminent and the cartel was in trouble. Its savior was the Ayatollah and "not the exemption of the market from the laws of supply and demand"), there does not appear to be a single trace of the collapse of the oil price in the period before the Iranian revolution, which was roughly the starting point for the second oil price shock. Not only was a 14.5 percent oil price increase scheduled in 1978 for 1979 (which was later telescoped into the first quarter of that year), but from the third quarter of 1975, the trend of the oil price on the Rotterdam spot market was unambiguously upward, with an unmistakable acceleration beginning in July 1978.

Can this recovery in the demand for oil take place again? If the world economy does not return to normality, and the aggregate (that is, macroeconomic) rate of growth in the OECD cannot be sustained at its present rate of 3 percent (which is also the level required to stop the increase in unemployment in the OECD), oil will tend to be in oversupply for the rest of this decade. However, eventually the upward pressure on oil demand caused by a sustained increase in economic activity in the industrial world, and the augmentation in demand that will accompany population growth in the Third World (even if per capita oil consumption is reduced below its already paltry value), will tend to offset those economic and political forces that have been active in curbing the demand for oil over the past few years. In addition, there will soon be a noticeable decline in the supply of oil from several of the most important producing areas outside the Persian Gulf (for example, the United Kingdom's North Sea). In the long run, if not sooner, the oil price will have to rise to keep the oil market in balance—and this is true despite the fact that at present less than half of OPEC's 37 Mbbl/d sustainable capacity is in use.

The internal discipline of OPEC is also important. During at least two periods that featured a sharply declining demand for oil (1974–1975 and 1982–1983), virtually all OPEC members cut back their production. This kind of self-restraint has surprised many OPEC watchers, although a simple explanation of these events might be formulated along the lines of enlightened self-interest—the same kind of enlightened self-interest cited by many economists in their explanation of why cartels inevitably break up. What the OPEC experience seems to reveal is that there is a certain aggregate production zone or band, perhaps between 16 and 17.5 Mbbl/d, where it is difficult for the members of that organization to cooperate (that is, agree on production quotas); but under 16 Mbbl/d they are driven to cooperation by a kind of desperation, and if the demand for OPEC oil is over 18 Mbbl/d the better-endowed OPEC countries can afford to be generous toward their poorer comrades and allow them to increase their production at a percentage rate that is higher than the average rate of increase for the organization as a whole.

At least four more topics must be presented before we leave this introductory section: the effect of the ending of the Iran-Iraq war on the supply of oil, the long-run effect of substitution and conservation on the demand for oil, the probable outcome of the future expansion of non-OPEC oil production on the oil market, and the possibility of politically motivated supply interruptions.

Taking the first of these items, it seems clear that after the termination of present hostilities both Iran and Iraq should experience little difficulty in recovering the greater part of their prewar production capacity; but no indication has been given by either party as to their production plans. However, since an important part of Iraq's war effort was financed by Saudi Arabia, it would be an act of distinguished ingratitude if Iraq chose to produce so much oil as to endanger Saudi Arabian oil revenues (and its own). It should also be remembered that the managers and economists of Iran are, in competence and sophistication, second to none in the entire oil-producing world and understand perfectly the laws of oil supply and demand. That they would deliberately spoil the market seems unlikely to this economist, although stranger things have happened on the upper echelons of international entrepreneurial activity in recent years.

As for the effect of conservation and substitution on the long-run demand for oil, the evidence is mixed, and I think that we will have to wait a few years before we know exactly which way the wind is blowing. My own approach to this issue centers on the amount of substitution and conservation that can take place in the industrial world before lifestyles begin to assume a different and perhaps unsatisfactory texture. I am not alluding here to the effect on three-car households of having to exchange a Cadillac for a Toyota, but to sizable changes in such things as employment and remuneration that may follow a drastic reduction in energy intensities—changes that, in my opinion, have already started to take place. Clearly, if such activities as the production of steel, aluminum refining, and a large slice of heavy manufacturing are dispensed with because they use too much energy, we are talking about fundamental and, perhaps, undesirable changes in the future prospects of our society.

Another point that should be brought up here is the rather embarassing correlation between the decline in labor productivity and the upsurge in energy saving investment. In the United States, productivity increases have reached a record low over the past few years, just as the adoption of energy-efficient technologies is moving faster than ever. This process is not very well understood, even though it is of crucial importance, and so I will provide a brief explanation. In a large part of the manufacturing (and commercial) sector, energy is replacing labor—in other words, many working places are being automated or robotized, and as a result individual factories or offices are becoming more rather than less energy-intensive. On the other hand, the people released from these working places move to less energy-intensive employment in the service sector, and concomitantly take a large reduction in income that reduces their access to energy-intensive activities (such as driving). Thus the entire society becomes less energy-intensive and, apparently, less productive. Since 1974, when energy input per unit of output first fell, productivity growth in the United States has averaged less than 1 percent a year, and as table 1–1 indicates, something has also gone very wrong in other parts of the world. Although it is too early to attempt a scientific appraisal of the situation, the effort to do without energy—rather than to alter the technology for supplying energy radically—may turn out to be a major

Table 1–1
Selected Performance Statistics for the OECD and OPEC Crude Oil Production, 1973–1983

Year	*OPEC Production (Mbbl/d)*	*Percent Change in Industrial Production, 7 Largest OECD Countries*	*Unemployment in 15 OECD Countries (percent)*	*Gross Investment: Deviation from Trend, 7 Largest OECD Countries*
1973	32.0	+6	3.20	+10%
1974	31.0	+0.5	3.75	+5%
1975	26.5	−8.5	5.20	−7%
1976	31.0	−2.5	5.25	−4%
1977	32.4	+1.0	5.25	0
1978	29.0	+2.0	5.20	+4%
1979	31.4	+5.0	5.05	+6%
1980	27.3	+2.2	5.95	0
1981	22.0	+2.1	6.90	0
1982	18.6	−5.0	8.20	−6%
1983	15.5	−12.5	8.50[a]	−12%[a]

Source: *Shell Briefing Service*, various issues: Bank for International Settlements, 54th Annual Report, 1984; *OECD Economic Outlook*, 1982, 1983, 1984.

[a] Estimated

economic and social blunder, merely hastening the displacement of a large part of the OECD's industry to East Asia. At the same time though, I see no reason to conceal my opinion that unless wages and salaries in North America and Western Europe can be kept in line with productivity, which may be a political impossibility, then that is where factories and workshops belong. As an aside let me mention that the wages and salaries I am mostly referring to are not those in the industrial sector. Taking Sweden as an example, the decline in competitiveness of Swedish exports from 1976 to the massive devaluation of 1982 was not, for the most part, the result of excessive rises in manufacturing wages and salaries, but rather caused by the rise in price of all Swedish products originating with the presence of too much purchasing power at the disposal of nonindustrial employees.

Non-OPEC oil production was almost unchanged between 1973 and 1976, but since that time has grown at about 6 percent annually, and in 1983 reached approximately 25 Mbbl/d. The North Sea and the North Slope of Alaska have been responsible for much of this gain, and Mexico has also made an important contribution. The great hope outside OPEC is, as to be expected, that Mexico will take steps to deplete its reserves at a faster rate, and therefore ensure that the excess-demand gap that many of us expect to eventually appear can be immediately closed.

My opinion, expressed at some length in *The Political Economy of*

Oil, is that Mexico would be doing many of us a favor if it squanders its precious hydrocarbon resources, but not the Mexican people. Unless I am mistaken, however, none of the major political parties in that country have plans to lift and sell Mexican oil at bargain basement prices, which is what they would be doing if, by the middle of this decade, Mexico raises its oil production as fast as several potential oil importers want it to. (The target once mentioned was 7.5 Mbbl/d, but this figure is seldom used any more.) Proven oil reserves in Mexico have increased from 11.2 billion barrels (11.2 Gbbl) in 1976 to more than 50 Gbbl in 1984, while oil production went from 897,000 bbl/d in 1976 to 2.7 Mbbl/d in 1983, when oil exports reached 1.5 Mbbl/d. The importance of oil to Mexico can be judged by a cursory examination of that country's national accounting. Oil represents 10.2 percent of Mexico's gross domestic product and 78.3 percent of its total merchandise exports. The oil industry also contributes about 37 percent of total tax income and accounts for more than one-third of public investment. In a country where population growth and unemployment are almost out of control, a premature exhaustion of oil reserves would be a simple prelude to disaster.

Other non-OPEC oil-producing regions of interest are the Norwegian North Sea and the Soviet Union. The policy of the Norwegians is, and always has been, to make sure that their oil reserves last at least 100 years, by which time these oil assets are to be replaced by other assets having (at least in theory) the same earning capacity. Most of the people I have talked to feel that this is the only sensible approach to the problem, although some seem disappointed when they find out that Saudi Arabia, Kuwait, and the United Arab Emirates have the same intentions. The Soviet Union has been exporting about 1.5 Mbbl/d of oil and oil products to the West over the past few years and perhaps the same amount to Eastern Europe. There is a theory that these exports cannot continue indefinitely and, in order to keep up the flow of hard currency that they bring in, the Soviet Union will make great efforts to increase its exports of natural gas. This is probably true; however, the Soviet Union consumes roughly 9 Mbbl/d of oil domestically, and a significant amount of this can probably be replaced with nuclear energy, gas, and coal. In fact, unless the Soviets lose their taste for foreign currencies, this replacement will have to take place, because although the market for natural gas is going to expand rapidly in the next decade, it is almost unthinkable that this market can provide the revenues for the Soviets that they now obtain from the export of oil.

Finally, we come to the matter of supply interruptions that have their basis in political phenomena. Whenever this topic is brought up it inevitably leads to a consideration of those factors which might affect the

Middle East and its large exports of oil. The Straits of Hormuz are generally considered to be the Achilles Heel of Middle Eastern oil, but the decisionmakers in Washington and Western Europe tacitly believe that nothing can go wrong in these waters that an aircraft carrier and a few battalions of marines or paratroopers cannot put right. On the other hand an insoluble problem might be created by a fundamentalist revolt in Saudi Arabia that turned off the oil spigot on religious or diffuse political grounds or by a general Soviet attack on the Gulf. As for the first of these events, I personally feel that things are proceeding in Saudi Arabia in such a way as to make a coup or revolution less rather than more likely. Saudi Arabia is one of the few less-developed countries in the entire world that has designed, and is following, an optimal development program. By the same token, it is extremely unlikely that the Soviets would risk nuclear disaster or the interruption of their growing trade with the capitalist world to seize energy materials they do not need.

Some Basic Petroleum Economics

The first point to remember here is that elementary economic theory almost always prescribes an increase in the supply of a commodity following a large increase in demand. The problem with the oil market, however, is that it does not function as depicted in most microeconomic textbooks, and more often than not its behavior appears to contradict some of the best known propositions of conventional economic theory. At this junction it might be useful to note that for the first time in the history of the United States, the drilling boom that followed the 1973–1974 oil price rises failed to increase 'lower 48' oil reserves. The reason for this is simply the evolving scarcity of onshore oil in the United States, which is a situation that markets can sometimes register but, when dealing with geological constraints, cannot always remedy. It is also striking that on a worldwide basis there has been a dramatic decline in the amount of oil discovered, despite the fact that more drilling is going on in more places than ever. In the early 1960s about 38 Gbbl of oil were being discovered every year, while by the late 1970s this figure was down to 10 Gbbl/year.

What about long-run supply, which is contingent on the amount of oil in the ground, and the technology for exploiting it? This is an interesting and important issue, and the reader should make an effort to follow the remainder of the discussion in this section, for there have been a great many serious misunderstandings about these matters.

At the beginning of the 1970s a consensus existed among leading petroleum geologists that the quantity of recoverable oil in the crust of the earth amounted to 2 trillion (or 2,000 Gbbl) barrels. Of this 2 trillion, 1,200 Gbbl has been located, and of these about 450 Gbbl consumed. Thus 800 Gbbl remains to be found. Exercises of this type still take place periodically, with both oil company and independent geologists asked to provide estimates of the total amount of recoverable oil that was originally in the earth's crust, using the latest information at their disposal. For the most part these estimates continue to cluster around 2 trillion barrels, even though the price of crude oil has increased by a factor of eight since the first of these exercises took place. However, in 1983 the United States Geological Survey (USGS) released Open-File Report 83-728 in which it was suggested that only about 1,718 Gbbl will eventually be recovered by conventional means. Undiscovered reserves have thus been reduced to about 520 Gbbl. Of greater significance, however, is the contention that there is strictly no possibility of another massive oil province being discovered, which implies that most of the new reserves will be in the Middle East.

The key term is *recoverable* oil, because not all oil in place is recoverable. In the United States, between 10 and 60 percent of the original oil in place can be recovered by either primary or secondary methods, with the overall recovery average being about 31 percent. Primary methods involve natural release or simple pumping, while secondary methods involve pumping water into the stratum below the oil deposit or gas into the layer above. The intention in either case is to flush out the crude oil clinging to the reservoir rock; and these two procedures constitute the major component of any oil-producing operation, even though their application still leaves, on the average, two barrels of oil in the ground for every barrel that is lifted.

But there are other processes that are relevant here, and these are generally called *tertiary* or *enhanced* recovery. The most frequently applied tertiary methods call for lowering the viscosity of oil in a reservoir so that it can flow more freely, usually by heating the oil (by injecting steam into it), or by injecting chemicals into the reservoir. Tertiary methods are used more liberally in the United States than anywhere else, and as a result the average recovery rate in the United States (about 32 percent of the oil in place) is slightly higher than in the rest of the world. It has been suggested that eventually the average U.S. recovery rate will reach 40 percent.

Unfortunately, no responsible person knows exactly what eventually means in this context, because in the past decade intensive and costly research has done everything possible to bring about a breakthrough in the technology for tertiary recovery without achieving other than some minor successes. In fact, as far as I have been able to tell, the average recovery rate in the United States has not increased by more than a fraction of 1 percent since the first oil price shock and may well be decreasing as the richer oil fields are depleted. Moreover, due to the recent stagnation in the oil price, a great deal of this kind of research is being terminated because, in the opinion of some of the most influential oil industry executives, tertiary methods will never be viable except under special conditions.

The intensive exploration and drilling that has been going on all over the world during the last decade has led to more successes being recorded than ever, and many of the new deposits being exploited are in relatively attractive locations, such as off the coast of West Africa or Brazil. The problem is that in terms of quantity, these deposits do not contain enough oil. What is needed to decisively augment world reserves of recoverable oil is for a few giant or supergiant fields to be discovered, because despite the fact that there are well in excess of 30,000 oil fields in the world today, almost 70 percent of world oil production (and 75 percent of world reserves) are accounted for by the 1 percent of these fields classified as giant or supergiant. (A giant field is one containing at least 500 Mbbl of oil, or 3 trillion cubic feet of natural gas.) But none of these giant or supergiant fields has been brought into production in more than a decade, despite the well-known fact that the economics of these fields is such that they always yield enormous profits to their owners. (For example, preparations to exploit the larger fields of the North Sea and Alaska were undertaken when the price of oil was less than one-tenth of that prevailing today.) In addition, new giant or supergiant fields are not being found despite vastly improved techniques for doing so. This is due to the fact that there are fewer of these fields left to be found. Most of the world's oil is concentrated in only a few basins and relatively few fields; and of the 600 or so bona fide basins in the world, 400 have been thoroughly explored, and only 160 have yielded commercial discoveries. As for those that have not been thoroughly explored, preliminary examination indicates that it would be unprofitable to do so at the present time. In line with the previous remarks, only 25 of the 400 basins have experienced discoveries of at least 10 billion barrels, and these 25 contain 86 percent of the total hydrocarbons discovered. Given the present state of the world economy, some people might argue that the failure to locate at least a new giant oil field every year or so can be overlooked; but if the world economy begins to function in such a way as to bring a halt to the present stagnation in growth and employment, the present trend in regard to the depletion of existing reserves will soon take on serious overtones.

Moving from the supply to the demand side, some attention needs to be paid to the proposition that the oil price shocks have permanently lowered the demand for crude. Certainly, demand recovered rapidly enough following the 1973 price shock, and a large part of this recovery can be attributed to a resumption in OECD income growth, which, although lower on the average than pre-1973 values, managed to exceed 3 percent a year for several years. In addition, there was a reversal during this period of the energy consumption–gross national product ratio, which had begun to fall in 1974. After the second oil price shock, with oil consumption sliding again, many analysts tended to overlook this period, and instead claimed that the decline in oil consumption in the 1980s was, in reality, just the delayed effect of the 1973–1974 oil price rises; and the full impact of the 1979–1980 oil price increases are yet to be felt.

As far as I can tell, this kind of opinion is incorrect. The low price elasticities of demand for oil preclude extremely strong delayed effects, and in the face of a sustained upswing in gross national product in the major oil-importing countries, the income elasticity of demand for oil is still large enough to lift consumption by a large amount. Probably the most important factor working to hold down oil demand is the increasing involvement of governments in the energy picture, since it is becoming a widely accepted truth in official circles that oil has a macroeconomic as well as a microeconomic dimension. In the factual climate of what might politely be called uncertainty, energy crises would arrive as frequently as elections if informed governments are unable to devise effective and durable policies for altering the natural course of energy demand.

Calculating the Price Elasticity of Energy

In the previous discussion, it was suggested that the *own* price elasticity of oil is fairly low. This is, quite naturally, an empirical question, and it is only fair to point out that some economists consider it to be fairly high. There are also some serious and, as yet, unresolved econometric problems associated with estimating own price elasticities for energy materials. When I taught econometrics the consensus view was that time series estimates of elasticities were superior to cross-section estimates; but it has been argued that because of the lack of variability in the oil price prior to 1973, as well as the comparative shortage of observations after 1973, time series estimates of the elasticities of oil and oil products are systematically biased downward. Instead it has been suggested that pooling international cross-sectional and time series data will yield more satisfactory results. Certainly the elasticities that are obtained employing the latter type of data seem to be twice as large as those obtained using time series data alone; but my intensive study of own elasticities from individual countries, which show a very wide range, leads me to question the value of any aggregate measure derived from data from economies having very different economic characteristics, particularly in regard to the structure of the economy and its stage of development. In these circumstances I prefer a simpler type of calculation, which I present here for energy. First, I define the demand for energy as a function of its price and some aggregate variable such as gross national product or industrial production. Thus we have $q = q(p,y)$, and a total differentiation with respect to price yields:

$$\frac{pdq}{qdp} = \frac{\delta q}{\delta p}\frac{p}{q} + \frac{\delta q}{\delta y}\frac{dy}{dp}\frac{p}{q}\frac{y}{q}\frac{q}{y} = n_p + n_i \frac{p}{y}\frac{dy}{dp} \tag{1.1}$$

Next, going over to finite movements in the variables, equation 1.1 can be written as:

$$\left(\frac{\Delta q}{q}\right)_j = \left(\frac{\Delta p}{p}\right)_j n_p + n_i\left(\frac{\Delta y}{y}\right)_j$$

or:

$$\hat{q}_j = n_p\hat{p}_j + n_i\hat{y}_j \tag{1.2}$$

In these expressions, n_p and n_i signify the partial elasticities of demand with respect to price and the aggregate variable y. Now, for $j =$ 0,1, where 0 and 1 are averages covering several years, we get:

$$n_p = \frac{(\hat{q}_o - \hat{q}_1) - n_i(\hat{y}_o - \hat{y}_1)}{\hat{p}_o - \hat{p}_1} \tag{1.3}$$

Let us employ the following data. The percentage growth in energy prices in the United States from 1965 to 1973 was −0.1, and from 1973 to 1978 it was 6.8. The percentage growth in energy demand is 4.3 for the 1965–1973 period and 0.9 for the 1973–1978 period. The percentage GNP growth for the 1965–1973 period was 3.7, while for 1973–1978 it was 2.3. Insofar as n_i is concerned, this was taken as unity. Substituting these values in equation 1.3 then gives:

$$n_p = \frac{(4.3 - 0.9) - 1.0\,(3.7 - 2.3)}{(-0.1 - 6.8)} = \frac{2.0}{-6.9} \approx -0.3$$

This estimate of the own elasticity of energy demand for the United States compares favorably with other estimates. It should also be pointed out that the absolute value of the price elasticity of oil would normally be larger than that for energy, since substitution between energy types is much easier than, for example, substituting labor or capital for energy.

OPEC and the Future

Before going to the most important topic concerning the world oil economy at the present time, which is the revolution in the refining sector, a few more things need to be said about the ongoing development of OPEC.

As has already been explained, in 1979 OPEC produced almost 31 Mbbl/d of crude oil and natural gas liquids, while today production is somewhere in the vicinity of 18 Mbbl/d, and on various occasions has been as low as 15.5 Mbbl/d. Moreover, it will be a very long time, if ever, before OPEC production reattains its 1979 value, given the amount of oil that is being produced outside of OPEC and the worldwide slump in the demand for all energy materials. What should not be forgotten is that the market for coal is now experiencing its worst slump since the 1960s; the demand for natural gas has dropped so rapidly that countries like Algeria, which at one time refused to discuss any price for gas under an oil parity price, are now willing to sell this commodity for a bargain basement price.

For some, if not all, OPEC countries the present situation is very unhealthy, but unhealthy in different ways. First, the loss of billions of dollars in oil revenues cannot possibly pose a fatal, or even near-fatal, threat to the economies of the richer OPEC countries. In fact, in many ways the fall in oil prices is a blessing for these countries because it focuses their attention on more substantial issues, such as the need to increase their output of refined products and petrochemicals and to increase the efficiency of their educational systems. Also, because of the weakening of the political and economic power of OPEC due to the recent oil surpluses, it has become much easier for various OPEC countries to purchase physical assets in the major oil-importing countries, including outlets for distributing their oil and oil products. The problem for this group of countries is that the worldwide fall in the demand for oil delays the day when, due to a visible depletion of their reserves, some of the less rational or more corrupt politicians in some OPEC countries have less to say about OPEC price and production policies.

Some of the weaker OPEC countries are in a position where they can literally see the exhaustion of their oil assets but, in the light of present and expected future oil prices, it appears that their oil will be gone well before they have entered into even the initial stages of self-sustaining economic growth. During the past few years some of these countries have been covertly experimenting with independent pricing arrangements that are more or less outside the general OPEC pricing guidelines. This, of course, is a losing game for these countries in a weak oil market, and had these experiments continued, the price of oil might well have spiraled downward, and the economies of these weaker countries ruined or at least badly damaged. The governments of most oil-exporting countries now give the impression of understanding how to price exhaustible resources, but in case they do not, it reduces to the following: Sell as little of the resource as possible at the highest possible price. To this can be added: Ignore the advice of half-baked academics about how essential it is to produce oil in order to pay inflated prices for show projects, fancy armaments, and other socially useless rubbish.

The final point in this section concerns the likelihood that the world economy will move into a full-scale recession again and, in the confusion that follows, the price of oil will begin an uncontrollable descent. The first thing that should be appreciated here is that there is a floor under the oil price that is set—for example, by the extraction costs of frontier or offshore oil—probably somewhere in the range of $15 to $20 per barrel. What cannot happen is that the oil price is free to descend to "any level," to use a specimen of the rather peculiar terminology that the economists of the Hudson Institute like to employ when discussing the future of the oil market with their clients. Even more important, the farther the price of oil falls and the longer it stays depressed, the more demand is eventually stimulated and the more rapidly oil prices can be expected to ascend. This notion is said to be very widespread in the executive suites of the larger oil companies, and as far as I can tell is at least partially understood by the OPEC directorate.

What does not seem to be understood by decisionmakers in all the OPEC countries is that the immediate future—that is, the next four or five years—is the most important phase of that organization's existence. If the wrong production and allocation decisions are made during that period, the result will be losses of hundreds of billions of dollars in present and future revenues from both oil and other economic activities. This will undoubtedly make certain people happy, but it will not change the fact that the future of the world oil market will still be decided by those countries controlling the most oil in the ground. Although production has been rising in an impressive manner outside of OPEC, and may be capable of continuing to do so for a few more years, the known stock of extractable oil in the ground is not rising. On the contrary, it may be falling, and with it the possibility that the oil-importing countries can call their energy future their own for more than another decade or so unless they reduce even further their dependence on this particular energy medium.

Refining: The Quiet Revolution

As an introduction to this subject, the international oil industry comprises both the majors (Exxon, Mobil, Texaco, Royal Dutch Shell, British Petroleum, SoCal-Gulf) and a large number of minors or independents, some of whom in terms of assets and sales would be considered majors in any business except petroleum.

These organizations have received some heavy punishment from fate (in the form of OPEC) during the past decade, and as will be argued later, more could come. In 1973 the majors and larger independents lifted 25 Mbbl/d from either their own properties in the oil-exporting countries or properties that could be regarded as their own; but by 1984 this figure was slightly under 7 Mbbl/d. Some of this equity (or near-equity) oil was replaced by purchases from the producing countries so that the oil companies could keep their refineries in operation, but over the decade beginning in 1973, 15 Mbbl/d was totally removed from the control of the international oil industry. This did not mean, incidentally, that all these firms became poorer. Because of the increase in the price of oil, those companies owning large reserves in stable areas like Alaska or the North Sea actually became richer, and one of these—Exxon—replaced General Motors as the number one private firm in the world. But the combination of confiscation by OPEC, high taxes, the public's disrespect, a slack market, and the difficulty in finding more exploitable reserves has taken some of the joy out of the lives of the directors of these organizations. Many of these gentlemen are now having to settle down to a no-growth situation for the first time in their lives.

The relative decline of the oil companies has been matched by the climb to prominence of the producing countries. They are, for one thing, dealing with larger numbers of buyers in the oil-importing countries, and they have also increased the amount of business they are doing on the spot market. There have also been some important changes in the duration of contracts. The so-called long-term contract has been decreasing in length, initially because the sellers of crude did not want to be committed to a near-stationary price during a period when they thought prices were due to explode upward, and lately because buyers of oil refuse to sign long-term contracts, since with a falling price they expect to be able to obtain their requirements cheaper in the future.

The full story of the trading in crude oil can be found in *The Political Economy of Oil,* but since that book does not treat downstream operations in great detail, I propose to concentrate on this matter here. One of the theories advanced in my earlier book was that the center of gravity of world refining is slowly shifting toward the large petroleum producing countries of the Middle East, and at the time this conjecture was bitterly opposed by various people in the academic and business world, many of whom seemed to feel that denying the feasibility of this move could prevent it from taking place.

The first step in the exposition is to describe a refinery. As shown in figure 1–1, a refinery is an installation for turning crude oil into a *slate* of various products. By convention, these products are divided into three cuts or fractions: gas and gasoline (light products), middle distillates, and fuel oil and residual cuts. At present the most valuable end of the barrel is the top end, which provides so-called white products: domestic gases, aviation fuels, motor fuels, and some feedstocks for the petrochemical industry. Naphtha, which is important for the improvement of gasoline quality and is also a valuable petrochemical input, is extracted from both the light and middle ranges of distillate cuts. The other middle distillates are kerosene, paraffin, and light gas-oil; the rest of the refinery output consists of heavy lubrication oils and residue.

This system functions as follows. Crude oil is pumped into a tall distillation tower that is pressurized and is hotter at the bottom than at the top. The various oil products have different boiling points, with those that are the lightest having the lowest. When the crude enters the tower, the heaviest part remains in liquid form and falls to the bottom. The rest is vaporized, but the various constituent products return to liquid form as they reach the lower temperatures higher up the column. As a result they can be piped away.

Products such as aviation gasoline, diesel oil, and waxes are produced from such basic products as naphtha and kerosene. Note also that the proportion of these products depends on whether the oil that is used as an input is relatively heavy or light. As shown in figure 1–1, a light oil produces a larger proportion of light products, while a heavy oil results in more heavy products. Oil is graded according to an API (American Petroleum Institute) number, and the higher the number, the lighter the barrel. It should also be appreciated the oils of different weights are generally located in different places. Venezuelan and Mexican oil tend, on the average, to be heavy, which in view of today's market preference for light products, causes them to sell at a discount to the lighter oils of, for example, Libya. Also, a country like Saudi Arabia possesses substan-

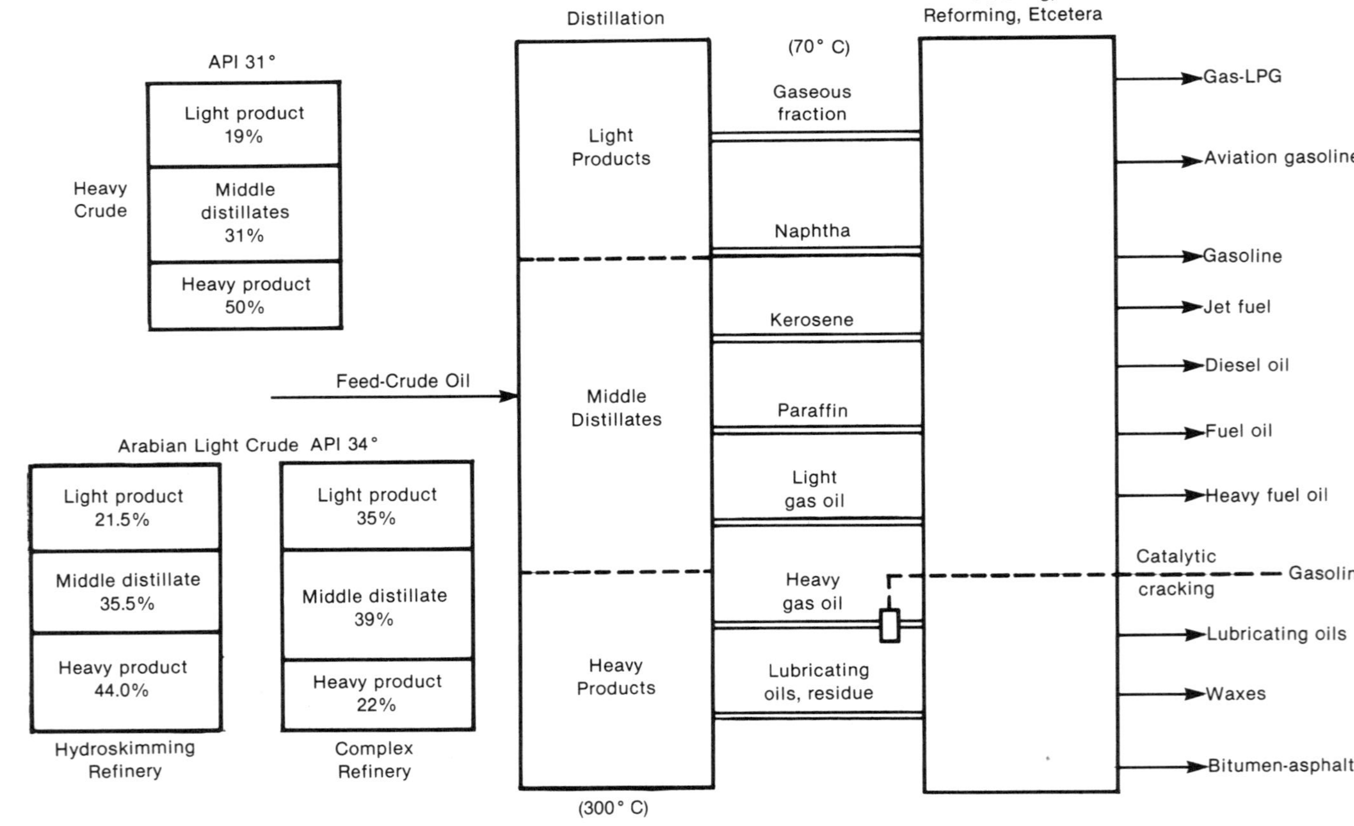

Figure 1–1. Input and Output Pattern in a Typical Refinery

tial quantities of both light and heavy oils, as well as considerable choice as to the mix of these it will produce.

The changing pattern of demand has led to a dramatic attempt to upgrade (or retrofit) refineries. With the yield from hydroskimming refineries (that is, refineries with atmospheric distilling, reforming, and hydrotreating plants) no longer conforming to the structure of demand, it has become profitable to alter output mixes by the addition to existing refineries of complex refining procedures, and in certain situations to build new refineries designed to provide a maximum output of high value white products. The process introduced to break down the heavier streams is known as *cracking;* and catalytic crackers are the basic type of equipment in this process. Another important process is known as *reforming*. Using heat and pressure, reforming changes the chemical structure of hydrocarbons in such a way as to produce new compounds. In 1979–1980 there were 289 refineries in the United States, with a combined capacity of 17.17 Mbbl/d. Catalytic cracking capacity came to 5.05 Mbbl/d and reforming 3.84 Mbbl/d. Unfortunately, putting in such things as catalytic crackers is an expensive proposition, particularly in the face of a growing surplus of refinery capacity and a fall in the demand for all oil products; but even so the larger oil firms are not hesitating to spend up to $1 billion to upgrade those refineries which they feel will help them to consolidate or expand their markets.

By way of contrast, one of the reasons so many independent refiners in the United States are going bankrupt is that they cannot afford to upgrade their refineries; however, even those who successfully upgraded their facilities often operate with a loss if they did not upgrade at the right time. In 1980 independents owned 183 refineries in the United States and produced about one-quarter of that country's oil products. More than 50 percent of these refineries have gone out of business, but the United States still probably has 25 percent more refinery capacity than is required. Similarly, 25 percent of the refinery capacity that existed in the European Economic Community (EEC) five years ago has been shut down, but there is still a huge excess capacity in Western Europe that will eventually result in many more refineries disappearing. In the face of competition from the Gulf, there is no other outcome, since complex refineries constructed in the oil-rich countries of that region can be supplied with inexpensive inputs which more than offset the expense of shipping products. (Oil products are shipped in smaller, more specialized tankers than those used for crude oil.)

It goes without saying that no one can foretell the long-run outcome of the refinery revolution, but three things seem likely. First, many independent refiners are on their last legs in the United States. Second, very many, but not all, European refineries are in a similar situation.

Third, the center of gravity of world refining is on its way toward the Gulf, although it cannot be expected to arrive in the immediate future.

To take the first of these points, small refineries in the United States require protection—for example, the entitlements program that required crude-rich refiners to subsidize those without crude. Since the government of President Reagan ostensibly does not believe in entitlements, these were removed with the expected results. In the same vein the American Independent Refiners' Association is requesting that the U.S. Congress put a tariff on imports, but hardly anyone even vaguely in touch with reality expects that such a measure will be approved in the near future. Furthermore, on the U.S. West Coast, falling oil prices have driven down gasoline prices, and in the struggle for market shares the independent refiners have pushed them down even more. This struggle has meant that even some very large independent refiners are in the danger zone, particularly those who invested heavily in upgrading only to have the bottom fall out of the gasoline market. On the other hand, the majors will probably try to increase their market shares, since with more modern (and profitable) refineries they have little to lose by invading the established markets of independents. Still, they are in no hurry, given the cash flow problems now being experienced by many of the independents. Although it is not certain, word is out that these problems can only increase, since a number of financial institutions in the United States (such as the Continental Illinois Bank) have made many dubious or bad loans to energy companies, and are now withdrawing their lines of credit. Finally, there is the matter of new environmental regulations that will require the amount of lead in gasoline to be reduced to 0.1 grams per gallon. This is expected to cost millions of dollars per refinery that, for the most part, the managers and owners of independent refineries do not possess and have no way of getting.

The Western European refinery sector has many problems, but two are critical: the nearness of the European market to the Middle East and to Russia (which also has ambitions to export more of its crude oil in product form), and the low level of protection against foreign imports. Several of the major oil companies, who as a group are generally in a position to offset their losses in refining by profits on their other business, have been retreating from Europe for years, while those who have decided to stay are investing huge amounts of money in upgrading their refineries. Whether this will pay off in the long run remains to be seen, because even the Sicilian refinery of ISAB SpA (Italy) has lost business although it is one of the most modern installations in Europe. There will always be locally owned refineries in Europe because a sizable share of Europe's refineries belong to government-owned companies that will be forced to stay open for political and strategic reasons; but without protection from

OPEC exports, many private refineries are doomed. OPEC Middle Eastern refineries will soon have 5 Mbbl/d of capacity in highly flexible installations, and once they are assured of an unimpaired access to the European market, it should be possible to substantially increase this amount in a relatively short time. What must be comprehended here is that some OPEC countries intend to partially regain with refining and petrochemicals what has been temporarily lost on the oil price, and if possible, in the very long run, ship only oil products.

While a great deal has been made of the significance of falling oil prices for OPEC exporters, the significance of rising losses in the refinery sector for the oil firms has been serenely overlooked. Perhaps this is just as well, because certainly there is no possibility of the refinery sector as a whole getting back to the 85 percent utilization rate that is still considered necessary for profitability until very many more refineries leave the scene. I would also like to assure interested persons that the OPEC capture of the non-North American refining industry will not take place tomorrow, or even the day after, since OPEC's planned share of the refinery capacity in the noncentrally planned world is only 12 percent for the year 1990. Excluding North America, however, OPEC expects to have 17.5 percent in that year, and from that point on it should be roses all the way. True, the rest of the world's refining plants will be more flexible because of the shake-out of this decade (flexible in the sense that refineries can handle any type of crude), but what they cannot handle are refineries in the Middle East—and perhaps also Libya—that can run on gas costing 50 cents per million British Thermal Units (50¢/MBtu). This happens to be only 10 percent of the average European fuel cost. Furthermore, if the price of crude oil does begin to rise in the 1990s, these Middle Eastern refineries will be obtaining their crude at the cost of taking it out of the ground, which at the most will be well under $5 a barrel.

Oil Futures Markets and Inventories

The next topic, futures markets, has become of increasing importance over the past few years. Only a brief review will be provided here, but Prast and Lax (1983) and Banks (1983) provide a more extensive exposition.

Futures markets function as follows. Against a background of speculators betting on the direction and size of commodity price movements by buying and selling futures contracts, an impersonal agency can be created that permits producers, consumers, inventory holders, and other traders in physical products to reduce (that is, hedge) undesired price risk. As simple as all this turns out to be in both theory and practice, there are a great many misunderstandings about these markets. No less a personality than Senator Alan Cranston of California, who at one time thought that he could be elected president of the United States in 1980, has distinguished himself by assuring his constituents that futures markets were responsible for the high and increasing price of agricultural products. As it happens, however, futures trading has decreased the volatility and level of agricultural prices in the United States because, by facilitating the reduction of price risk, this trading encouraged producers and middlemen to carry larger inventories into the nonharvesting season. In the case of oil and oil products, the advantages of futures trading may eventually include less erratic price movements; but they will definitely feature a less ambiguous price setting and an upgrading in the efficiency of oil and oil product markets due to more information being made available to existing and potential market participants.

The success of a future market is dependent on the satisfaction of several well-defined criteria. It is essential that the commodity in question can be traded in bulk, is susceptible to grading, is relatively imperishable, and, most important, is bought and sold in circumstances that cause its price to fluctuate in a random or nonsystematic manner. Without this latter provision, speculators will not be attracted to the commodity, and without fairly large-scale speculation, futures markets will not function properly. Put another way, traders in a physical commodity can employ futures markets to reduce price risk only if other traders or speculators are wiling to accept this risk. The social gain from futures trading derives from the voluntary redistribution of risk from risk-averse dealers in physical products to speculators.

By the same token, very many speculators tend to lose interest in futures markets when they are confronted with stable prices. There were two failures of futures contracts for petroleum and petroleum products in the mid-1970s because of the return of the world petroleum market to a facsimile of normality. This hiatus, however, was only temporary and when wild price swings reappeared with the Iranian revolution and the war between Iran and Iraq, there was a sharp escalation in petroleum futures trading in New York (on the New York Mercantile Exchange) and, eventually, on the recently established International Petroleum Exchange in London.

Before continuing, an important detail deserves some attention. The futures markets being considered here are mostly for oil products—principally gas-oil (that is, No. 2 heating oil) and motor gasoline. Contracts have just become available in England and the United States for crude oil, and apparently there are plans for reintroducing a contract for No. 6 fuel oil, as well as talk about launching contracts for other high volume petroleum products, natural gas, and coal. The key thing, though, for hedgers (that is, risk-averse buyers and sellers of a physical product) is that the prices of those items for which contracts exist are highly correlated with the prices of other petroleum products and crude oil. If this is the case, then price risk associated with one or more commodities can be reduced by purchasing or selling futures contracts for gas-oil. (This type of arrangement is explained later.) There is, however, a sound argument for introducing additional contracts. Further contracts are likely to increase the volume of speculation, which means that a larger amount of hedging can be accommodated.

Now we can turn to the behavior of hedgers, although first the reader should consider a few remarks about the modus operandi of speculators. If a speculator believes that the price of a commodity is going to rise, he buys futures contracts. These contracts are also forward contracts, in that delivery conditions are stipulated on them relating to a specified amount of a commodity, delivered during a certain month to one or more specified locations; but it is possible to avoid taking delivery if, at any time before the contract matures, an offsetting sale is made of a contract for the same amount of the same commodity, referred to the same delivery month. Offsetting, or "closing a position," is always possible in a viable futures market. If the price at which the contract is sold is higher than that at which it was bought (that is, the price at which the position was opened), the speculator has made a profit. Similarly, if the speculator thinks that the price is going to fall, he opens his position by selling a contract, hoping to make a profit by closing his position at a lower price.

Hedging

Hedgers also buy or sell futures contracts, depending upon whether they want to guard against price rises or price falls. Consider, for example, someone who has contracted for a given quantity of heating oil, but who does not know the price at which the oil will be delivered because oil companies make a practice of charging the posted (that is, spot) price for oil at the time of delivery. This buyer thus faces considerable price risk in that the price of oil might rise sharply. (But notice that even if it were possible to fix a price in advance, price risk is still present. If the price of the oil falls, those who have bought forward will be at a competitive disadvantage to those who waited and bought spot.) Risk-averting buyers can therefore lock in the price at which they will receive their supplies if they buy futures contracts at the time they contract for their oil, while making an offsetting sale around the time the oil is delivered. If the spot price of the oil rises, the buyer takes a loss on the physical transaction;

however, if the price of futures contracts also rise in phase with the rise in the price of the physical commodity (which should happen), a compensating gain will be made on the sale of the futures. A perfect hedge is when the loss on the physical transaction is equal to the gain on the paper transaction—or vice versa. Perfect hedges rarely happen on individual transaction; but over a large number of transactions the small gains and losses per transaction should balance out. Hedgers will, however, have to pay a comparatively small brokerage charge for each transaction, which, conceptually, can be taken as analogous to an insurance fee.

What about the firm selling oil? The spectrum of petroleum futures traders is large, and includes refiners, terminal operators, jobbers, large consumers, stockists, and a few large oil companies. It seems to be true, though, that many large oil firms feel that they have enough experience in managing risk to ignore futures markets, and the not inconsiderable leverage of very large oil companies enables them to obtain or dispose of supplies at short notice. But even so, some of the larger organizations have begun to find access to the futures markets useful.

What needs to be understood here is that large oil companies have control over a certain amount of oil during a given period, and if this oil is not contracted for by their customers, then it is sold on spot markets where, ex ante, prices are unpredictable. But if these companies sell futures contracts, and the price of the commodity they are selling falls, then if the price of the futures contracts also falls—which it should—an offsetting purchase of contracts can be made at a lower price than the selling price. Consequently, the profit realized on the paper transaction tends to counterbalance the loss resulting from the fall in the spot price. Moreover, the presence of a futures market often allows a firm selling oil or oil products to quote firm (that is, fixed) prices to its customers. Cases are already known where large orders have been won in competitive bidding by sellers offering fixed prices instead of arrangements where the price depends upon aberrant and perhaps transitory occurrences on the world oil market.

A few issues remain to be cleared up. The most important of these concerns the convergence of the futures price with the spot price, together with the matter of deliverability. Formally, the criterion that has been proposed is that in its delivery month the price of a futures contract for any given commodity must be very close to the spot price of the commodity in the physical (or actuals) market. If this were not so, since delivery can be made on the contract, any discrepancy between the two prices would result in an arbitrage operation. Suppose, for instance, that the price of a futures contract was $1,000 on the last day of the month before the delivery month, while the spot price was $1,050. The buyer of a futures contract, perceiving this situation, does not make an offsetting sale, but accepts delivery on the commodity and immediately sells it on the spot market. This yields a profit of $50. The eventual prevailing of the law of one price will, of course, soon wipe out these opportunities for profit, precisely as it does in the foreign exchange market: the supply of the commodity increases and its price falls.

It should be appreciated, however, that in existing markets minor price discrepancies are common, even in the case of single grade contracts. This is so because delivery is not to the home of the buyer of a contract or to a local warehouse, but to specially designated warehouses, sometimes in out of the way places. In addition, delivery takes place during the delivery month, not on a specific date. Thus there is some uncertainty for the buyer as to the exact date and place of delivery, and unless the potential arbitrage gain is appreciable, many buyers will not bother to accept delivery. The same logic prevails for multiple-grade futures contracts, and in addition buyers taking delivery cannot even be sure of the exact product they are receiving.

The inconvenience of delivery has been one of the main factors causing only a few percent of the futures positions opened to conclude in deliveries; but it also happens to be true that more interest is shown in deliveries on petroleum futures than any of the other commodities. This is because of the great opportunity for substantial windfall gains that can occasionally arise in the oil and oil products market due to political and economic considerations, where the latter includes such apparent trivialities as differences in the weather conditions in geographically separate regions of the same country.

The present success of futures contracts in New York and London presages an expansion of these activities. Some question exists, however, as to the part the major oil companies will play in the futures trading drama. With the exception of British Petroleum, which was active in setting up the International Petroleum Exchange in London, many of the larger companies are still passive. A reason for this might be that the management of these companies are aware that a very large increase in hedging may not be possible without substantial price movements on futures markets, and such price movements often offend politicians, the media, the general public, and other market participants. The only defense against exaggerated price movements would seem to be a quantum increase in the speculative base of this market, including the introduction of new contracts, such as the new contract for crude oil.

Both major consumers of oil and oil products and the smaller sellers desire the protection that viable hedging facilities can offer, and if an upturn in world economic growth once again initiates an upward spiral in oil prices, it will be to everyone's advantage if these facilities are available.

Oil Stocks and Oil Prices

We know that where short-run pricing on many energy and nonfuel mineral markets is concerned, the level of inventories in relation to current consumption is the key variable. For example, it is easy to understand that on markets where stocks are very large, even strong surges in demand may not lead to appreciable price increases. When demand rises rapidly in elementary supply-demand models, prices increase because producers need higher revenues to cover the costs associated with a hurried expansion in output. But if large stocks are available, perhaps demand can be satisfied by just reducing the level of inventories. In these circumstances, at least in theory, the price of the commodity does not need to move.

Unfortunately, however, many aspects of this topic are not so straightforward, and so in what follows I attempt to provide a simplified exposition of some of these important matters. But before starting I would like to emphasize that many of the problems associated with expectations have not been solved, nor can they be solved within the ambit of mainstream economic theory, despite the impressions that might be given by a casual scrutiny of the learned journals.

Between the end of World War II and the oil price rises of 1973–1974, global oil inventories could be realistically estimated as a fairly stable percentage of annual oil consumption, generally amounting to 60 or 70 days coverage. Unexpected jumps in demand meant a sharp fall in stocks, but as a rule did not lead to frenetic buying to rebuild stocks (and thus to an upward pressure on the oil price) because it was generally felt that the balance of power in the world oil market was, and would remain, on the side of buyers. Similarly, a noticeable fall in the demand for oil did not lead to a precipitous unloading of stocks in anticipation of a fall in oil prices, which might have brought about a realization of that anticipation. This was because while a decline in the demand for oil was associated with a cyclic downturn of the world economy, the long-term macroeconomic growth trend was believed to be strong enough to maintain a steady rise in the world demand for oil, which in turn would support oil prices—in nominal (that is, money) if not in real terms. In addition, trading in oil took place largely on the basis of medium- and long-term contracts. The spot market was a relatively minor affair, and although any number of spectacular deals and coups are deeply embedded in the mythology of the world oil scene, there was not an unlimited amount of scope for ad hoc changes in behavior by large buyers and sellers. By the same token, movements in oil prices were so predictable as to make speculation in oil futures an unrewarding pursuit; and although oil was (and is) the world's largest business, futures traders tended to confine their efforts to buying and selling contracts for more prosaic commodities.

All this began changing in the first months of the 1980s. True, inventories had increased in the middle of the 1970s, mostly as a result of governmental stockbuilding; and there was a short-lived burst of activity in oil and oil product futures trading about the same time, but for the most part the international adjustment to the first oil price shock was completed by 1978, with income in the OECD rising by more than 3 percent a year and world oil consumption once more at the level of 1973. In 1979–1980, however, following the Iranian revolution, aggregate OPEC oil production began to fall, and with unchanged expectations about the growth of income and production in the oil-importing countries, it was widely predicted that oil prices would be higher and less stable. Accordingly, in late 1979, inventory holders initiated a steady accumulation of oil and oil product assets that, for all practical purposes, continued until late 1981 or early 1982. By that time, given the average consumption of the period, stocks had been increased to a coverage of well over 100 days. The excess demand caused by this abrupt departure from orthodoxy by stockists resulted in a drastic increase in the oil price that, affecting the world economy, brought about the deep recession that still characterizes a large part of the OECD. The exact situation with regard to stocks can be seen in figure 1–2, where it should be noted that days of consumption is an exact proxy for the consumption-inventory ratio just alluded to, and of crucial importance in the theoretical and empirical literature.

The recession speedily reversed expectations about future oil prices, and as a result the level of inventories was now too high, rather than too low. Thus a modest stock drawdown began in 1982 that eventually reached several million barrels per day. The result of this stock adjustment behavior was a sustained depression of the spot price of oil, since many of these stocks had to be disposed of through the Rotterdam spot market. Later on, of course, this deterioration in the spot price was reflected in quotations of the official price of oil—that is, the average price at which large OPEC members make their sales. Something that should be clarified at this time is that most of these stocks were not disposed of in a physical sense—that is, they were not consumed in current production or consumption processes—but rather they changed

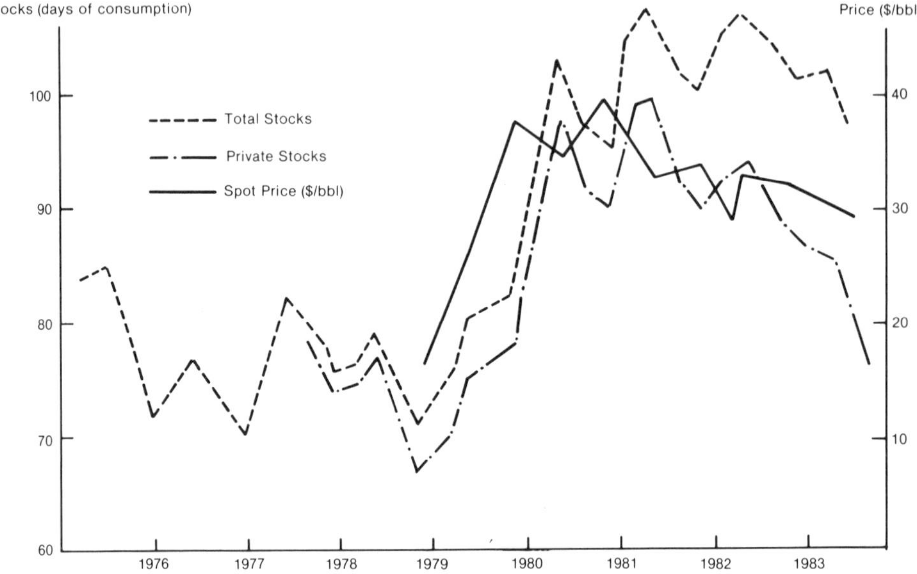

Figure 1–2. OECD Inventory Coverage in Days of Forward Consumption

owners, with their prices falling in the process. Put another way, given the expectations of future prices prevailing at that time, oil prices had to decline in order to justify the holding of existing stocks.

Inventories and Futures Markets

In 1980 the circumstances that make a futures market attractive to speculators appeared in full flower, and they are still in evidence to a considerable extent. By that I mean the increased variability in the oil price that began, it appears, with such things as the onset of economic stagnation in the world economy, uncertainty over the cohesiveness of OPEC, and other events. Furthermore, as the futures markets for oil products (and later crude oil) began to expand, they reinforced the inclination toward fewer long-term transactions in physical commodities. With an increased probability of falling oil prices, even fairly large buyers of oil and oil products knew that they could lock in the oil price at any given time by a judicious use of the futures market, and so they negotiated a smaller amount of their requirements through the medium of long-term contracts.

Analogously, there was a decreased propensity by some categories of buyers (for example, retailers of heating oil) to hold inventories as a precaution against having to buy in a rising market. Instead, individuals and firms requiring oil in the future bought a number of futures contracts. Later, they went into the spot market to buy their oil, while at the same time making an offsetting sale of futures contracts. If the spot price of oil had risen, they lost on the physical transaction but, as indicated earlier, they made a compensating gain on the sale of their futures contracts. Thus an obvious consequence of being able to resort to the futures market is a decreased need for inventories, along with a palpable increase in the number of spot transactions as a percentage of all transactions.

As indicated earlier, futures markets upgrade market efficiency by providing a powerful increase in the quality and quantity of information about a commodity. Consequently, markets with a large number of participants having varying degrees of risk aversion and many opinions about future trends or tendencies can be quickly brought into equilibrium. Just as important, expectations concerning the future performance of influential variables are continuously appraised and discounted. As a result, the prices displayed on futures markets are instrumental in forming expectations about future prices (that is, prices in the future) on physical markets, and these expected prices are essential for inventory holders trying to make what they consider to be optimal decisions.

Now that we have looked at the benefits accruing from futures markets, something should be said about the costs. It can be easily shown that the information provided by a futures market might have a destabilizing effect on spot prices due to its causing rapid shifts in the demand for stocks. There is no secret as to why this might turn out to be so, since if inventory behavior is influenced by changes in expectations about future prices, any institution that encourages frequent revisions of expectations can be identified as a potential source of increased price volatility. Just what weight should be attached to this defect is uncertain to this economist, although if mostly short-run prices were affected, it could be argued that what appears to be a shortcoming is actually a blessing. This is so because fluctuating prices attract a larger amount of speculation to futures markets, which by extension makes it possible to support an increased quantity of hedging (that is, insurance against price risk).

Finally, it should be remembered that governments are now involved in extensive stockholding and in the formulation of stockholding policies.

Members of the European Economic Community (EEC) have committed themselves to a plan worked out by the International Energy Agency (IEA) that involves holding stocks equivalent to 90 days of forward consumption. The only exception is the United Kingdom, whose obligation is 76.5 days because of its large domestic production and reserves. In addition, as shown in figure 1–2, governments now own an appreciably larger percentage of OECD stocks than was the case five years ago. Without going very deeply into the matter I can note that public inventory behavior, which keeps very large stocks in existence in relation to current consumption, is capable of engendering expectational patterns during a crisis that inhibit price rises and help to dampen price instability. What these stocks cannot do, however, is to compensate for the actual shortage of oil in the ground. Only sensible conservation and substitution measures, and cooperation between the producers and consumers of oil, are capable of bringing this about.

Oil and Macroeconomics

In the ten years that have passed since the first oil price shock, the world economy has taken the initial steps in the transition toward a new economic reality. Economic growth is down, unemployment is up, and prospects for the indefinitely rising living standards that, only a decade ago, seemed to be an integral part of the North American, Western European, and Japanese way of life have now diminished considerably. Although there is apparently a plentiful flow of oil to the main oil-importing countries at the present time, I am still prepared to argue that, at least in a limited sense, the eneregy crisis has not departed. This is because it is still impossible to obtain enough low to moderately priced energy to sustain (though not, perhaps, to temporarily reach) rates of economic growth capable of curing the macroeconomic agonies that have accompanied the fall in the consumption of oil. Put another way, the energy crisis as such has been replaced by an employment crisis.

To understand these issues better, let us take a look at what was happening in the industrial world from about 1950. To begin, the real price of oil declined by a factor of almost two between 1950 and 1970, although its money price was nearly constant. This decline made possible a sharp increase in the energy-intensiveness of productive activities in all the industrial countries, which in turn led to a dramatic rise in output in such sectors as aluminum, plastics, and steel. There was also a remarkable growth in the demand for and supply of private transportation. Even more important, energy-using technical progress brought about a large and durable increase in labor productivity. This rise in labor productivity was soon transformed into sizable real wage and salary increases for a large majority of the working population in the OECD, and subsequently into the high prosperity that characterized many parts of the world during the 1960s and the first part of the 1970s.

The oil price rises of 1973–1974 changed all this, even though the ensuing adjustment was not the disaster many of us thought it would be. For instance, with energy of all kinds providing for a maximum of 5 or 6 percent of the inputs into the productive activities of OECD countries (in terms of value), the approximately 400 percent oil price increases from October 1973 to the end of 1978 added 20 to 25 percent to the rise in the consumer price index over this period. Although modest by some standards, this is still a considerable amount when considered in terms of post–World War II history, for both traditionally low-inflation countries like the United States and countries like Sweden where the determination of money wages often tends to be associated with movements in the price level. Thus, in Sweden, the sudden acceleration in prices meant that the link between compensation and productivity was severely stressed, if not broken. In fact, if the reader is curious why a country like Sweden, which hardly experienced a large foreign debt or serious unemployment prior to 1973, is now plagued by both, the reason can be found in the attempt to maintain real purchasing power for all elements of the community in a situation where annual payments for energy imports increased by a factor of about ten.

At the same time the increased rate of inflation brought a benefit to the industrial world. It decreased the real price of oil in the sense that a barrel of oil cost less in terms of the exports of the oil-importing countries. Quite naturally, this process could not continue indefinitely, because eventually the oil-exporting countries would find it unacceptable and, if possible, raise the price of oil at an accelerated rate. But as far as I can tell, it was not until the late 1970s that the OPEC directorate was fully cognizant of the effect of the rate of inflation in the industrial countries on their own purchasing power. Moreover, as a result of differential rates of currency appreciation or depreciation, some oil-importing countries gained considerably more than others. The price of oil to West Germany, for example, fell steadily in relation to that of the United States, as the Deutschmark appreciated relative to the dollar, which is the currency used to pay for oil. (Analogously, the present disinflation in the OECD, together with the appreciation of the dollar, has helped to maintain the real purchasing power of oil, despite the weakness of the oil price.)

Concomitantly, the richer OPEC countries eventually recognized that they were not being provided with an adequate real rate of interest on their financial placements in the OECD. In 1976–1977, the real rate of interest (that is, the money rate of interest minus the inflation rate) was, on the average, very close to zero on bank accounts, Eurobonds, and so forth. After some very complicated negotiations and manipulations, these countries finally were able to make some very gratifying arrangements for their financial surpluses. Among other things, they were permitted to purchase some attractive properties in the industrial world, and eventually Saudi Arabia could greatly increase its influence in the International Monetary Fund (IMF). It was also during this period that the new industrializing countries (such as Taiwan and South Korea), which feature both low wages and increasing technical skills, were able to borrow money to build the steel mills and high-tech industries that will enable them to become full-fledged members of the industrial world in the not too distant future. Although it is not generally appreciated, this will not be good news for the present members of that fairly exclusive club.

In this tapestry of winners and losers, the oil-importing developing countries stand out as the biggest losers; but on balance the industrial world was also a heavy loser. Furthermore, the source of some of the most devastating losses to the international economy are still not adequately understood. The increase in oil prices resulted in the governments of energy-exporting countries like Britain seeking their salvation in oil revenues, instead of rebuilding the productivity of their industrial sectors by investing in a more modern production apparatus and establishing a more effective educational system (of the type introduced in Austria by the Kreisky government). Equally as alarming, the oil-importing countries put themselves into the position of having to adjust, piecemeal, to large balance of payments deficits. What this did was to fundamentally alter the momentum of world trade and, when the second oil price shock arrived, caused the volume of international trade to decrease for the first time in post–World War II history.

There were also serious adjustment problems associated with shifts in demand for both products and factors that deserve to be examined here. To begin with, consider an arrangement in which oil-exporting countries (OEC) spend their revenues in oil-importing countries (OIC) on a mix of goods and services that is very different from that which was being produced and consumed by the OICs prior to a large rise in oil prices. This state of affairs could lead to inflation or unemployment, depending upon the ease or difficulty with which factors of production can be shifted from one industry to another. Suppose that the residents of OIC normally prefer the goods provided by industry A, but when the OEC buys in OIC (with revenues earned from selling oil to the OIC), they desire goods from industry B. In these circumstances an oil price rise could lead to a lowered demand for A goods, and an increased demand for B goods; but prices in the real world go up more easily than they come down, and so we would expect price increases in B type industries, but hardly any movement in A type industries. Accordingly, the net result would be a rise in the general price level, and perhaps even a large rise. Moreover, there could be difficulties in reallocating factors of production from one industry to the other. Many of the employees, and some of the fixed production factors, located in industry A might no longer be needed in A or useful to B. And even if these employees were needed in B, and theoretically could be transferred immediately, rigidities in the wage structure might prevent employers in B from recruiting these people or, conversely, might not provide an incentive for them to move if it involved such things as changing localities and occupational preferences. The outcome of these events might therefore be involuntary or voluntary unemployment.

A number of severe complications were also introduced into the international financial system by having to recycle the balance of trade surpluses of the larger oil-producing countries. This recycling follows logically from the high oil consumption of the industrial world and the inability of the richest oil-producing countries to spend all the money earned by exporting oil within a short time of receiving it. Thus if surplus petrodollars could not be placed in various financial institutions at a satisfactory rate of interest, it would make more sense for the oil-exporting countries to leave their oil in the ground. Unfortunately, however, a satisfactory rate of interest is not easily obtainable in a world of rapidly increasing oil prices. This is so because many very large potential investment projects had to be cancelled because of the downturn in the world economy caused by these oil price rises and the increased cost of energy inputs for these projects. All else equal, the cancelling of these projects meant a decrease in the demand for loans, and thus an additional downward pressure on both the nominal (that is, money) and the real rate of interest.

Since high-class loans could not be made to the extent required to absorb the growing flood of petrodollars, the banking system eventually resorted to granting loans that were medium class or lower. Industrial countries like Sweden were able to borrow for consumption purposes, while increasing amounts of petrodollars were placed in the Third World. For example, Mexico borrowed in order to increase its oil production, while Brazil took large loans for the purpose of increasing its iron ore–exporting capacity. But after the second oil price shock, which generated many more petrodollars than the first oil price shock, the world was thrown into a recession in which less oil and iron ore was needed than ever. In these circumstances, the enormously expensive Mexican and Brazilian investments turned out to be highly unprofitable. The international economy is now in the process of adjusting to the lack of profitability of these and many other enterprises in the Third World, to falling energy intensities in the industrial world, and perhaps to a changing global income distribution. Over a long period this is going to mean, on the average, a slowing down in the rhythm of aggregate economic activity, and lower factor prices—in particular, real wages and salaries.

This is the background to the so-called debt problem of non-OPEC LDCs. In January 1984 their debt came to $560 billion, with two-thirds denominated in dollars, and at least a half bearing floating interest rates. Because the borrowing of petrodollars did not take place directly from OPEC, but was intermediated by some of the largest financial institutions in the world, everyone is now concerned with the possibility of a serious accident that, at least in theory, could disable the banking system and also affect the stock market. As things now stand, there are two likely outcomes to this predicament, both of them disagreeable: first, a repudiation or default (that is, a failure to comply with the terms of a loan) leading to a gigantic write-off or write-down by some of the largest banks in the world; and second, only slightly more acceptable, a transfer of the management of this debt to the taxpayers of the industrial countries by way of their central banks and with the blessing of their governments. The question then is whether this latter option can take place before a series of balance sheet rectifications must take place in the worst possible circumstances, and the international financial system is subjected to an intolerable strain.

In considering this matter, it should be understood that a voluntary write-off of a large chunk of sovereign debt in the name of international solidarity, or good public relations, is unlikely. The stockholders of the banks holding this debt would hardly be willing to stand by and see their equity vanish in front of their eyes, and the directors of these banks are well aware of this fact. They are also aware that many of their depositors would balk at keeping their money in institutions that are capable of this kind of behavior, and bank directors who lose the confidence of both stockholders and depositors do not have much future in this line of work. An event with even more potential for catastrope would be a write-off triggered by an unexpected moratorium called by several of the larger debtor countries, because in these circumstances the more sophisticated depositors of the banks holding this debt would immediately understand that it was in their best interests to withdraw their deposits as soon as possible. This is so because the equity or capital of many major banks is insufficient to cover even half their loans to only the four largest debtor countries. Thus, in the event these debtors cease payment, at least some of the deposits in these banks could be lost (unless these establishments can borrow on a long-term basis), and rational depositors will take immediate steps to make sure that they are not their deposits.

There is a strong belief now that central banks will fill the role of lenders of last resort—which actually means making good the losses of private banks. Since they were obviously prepared to come to the rescue of Continental Illinois, it seems unlikely that they would refuse to play the good samaritan in similar situations. It should be understood, however, that letting things deteriorate to a point where a lender of last resort is required is a suboptimal arrangement from a social point of view, since it will most likely involve the creation of liabilities backed by only bookkeeping assets (which happen to be called cash). The real asset, which is the ability of the debtor to repay the debt in goods and services, disappears when the moratorium or repudiation becomes final. The seriousness of the dilemma can perhaps be appraised from table 1–2, although in the light of these figures it is not easy to discuss, much less specify, an optimal policy. A free market solution would probably turn on lengthening the debt and spreading the risk of delay or loss over a very large number of stockholders, where these stockholders would be private individuals willing to bet a small amount that debts would eventually be paid and that they could purchase (as well as sell) debt in a secondary market. This assumes, of course, that such a market can be made to work, because attempting to unload only a portion of LDC debt on this kind of market today would depress its price by enough to threaten several large banks in Britain and the United States with insolvency. On the other hand, given the possible effects of an international financial crisis on the entire community, it might be argued that if private speculators are not available, the entire community should accept

Table 1–2
Third World Private Debt and Major Lenders to Brazil, Mexico, Argentina, and Venezuela

Country	*Debt (billion)*	*Bank*	*Loans/Equity (%)*
Brazil	$93	Manufacturers Hanover	240
Mexico	$89	Lloyds Bank	228
Argentina	$44	Midland Bank	213
Venezuela	$34	Chase Manhattan	175
Indonesia	$25	Citicorp	170
Philippines	$25	Chemical	165
Chile	$18	Bankers Trust	150
Turkey	$17	Bank of America	145

Source: Monthly newsletters, Morgan Guaranty; *OECD Economic Outlook*, 1983, 1984.
Note: Loans/Equity represents the ratio of bank loans to bank equity.

the risk via an agent—such as the central bank. With the complications that financial disorder might create for stockholders in nonbank corporations, for individuals who are concerned with the yield of pension funds, for depositors of private banks, and so on, there are almost certainly more people in the community who would approve rather than reject this course of action if they understood its ramifications.

As usual, there is no shortage of ivory tower proposals or even some nonproposals for dealing with the debt crisis at the international level. For instance, several years ago, in a speech at the Institute of Economic Affairs, in London, Milton Friedman expressed the belief that in the long run, if nature is allowed to take its course, everything will work out all right and, most important, nobody will find themselves on the short end. Certainly, Nobel Prize winners whose high-priced advice has left devastation in its wake from one end of the world to the other do not need to lose any sleep over what appears to them to be minor cracks in the international credit system. By way of contrast, the General Agreement on Tariffs and Treaties (GATT) is in favor of the worldwide promotion of investments in such things as metal processing and manufacturing, so that the economic activities indulged in by less developed countries would take on a new aura of profitability and thus eventually provide them with the means to pay their debts. Given the time factor involved, and the protectionism proposed by politicians, this so-called solution features the lack of comprehension that GATT's research elite has specialized in since the founding of that organization.

The New York investment banker Felix Rohatyn wants a new Reconstruction Finance Corporation. This organization would be capitalized at $5 billion (with an additional $25 billion borrowing authority) to handle problems concerning the United States. On the international scene he calls for a conference of creditors to deal with Third World debt. The outcome of such a conference would most likely be a setting of interest rate ceilings that would provide considerable relief for those creditors who have issued floating rate liabilities, as well as a farming out of selected portions of LDC debt to central banks and international organizations (for example, the International Monetary Fund). Everything considered, this might be the best possible alternative.

Since even Dr. Henry Kissinger now says that the debt crisis will not be dispelled by repayment—which is an insight undoubtedly gained from his many acquaintances in banking and government—that contingency was not discussed; but economic theory suggests that as soon as this reality is assimilated by the market, bank shares may fall in value to a point where they will be attractive to a less risk-averse group of investors who, among other things, would not panic at the first sign of trouble. As for depositors, it has already been suggested that in menacing situations there is a limit to their passivity. A widely publicized bank crash or near crash could initiate a chain reaction beginning with a massive transfer by individuals and organizations out of deposits and into what they believed to be superior stores of value, such as cash and physical assets. The effect of this kind of behavior could be a credit squeeze of draconic proportions that would be traumatic for countries and for many large firms trying to renew existing loans. It is this prospect that has not only bankers but heads of state and their advisers reaching for aspirin and tranquilizers.

Rumor has it that recently the U.S. Central Bank (or the Federal Reserve System) was considering buying the dubious loans to LDCs now on the books of private financial institutions, paying 90 percent of their value, with the form of payment being government bonds. The simple truth is that this should have been done a year ago, in a rising stock market, when investors in bank shares would have been more sanguine about accepting the 10 percent loss implicit in the transaction. At the time this is being written the stock market is falling, and given the bad news circulating on Wall Street these days, a 10 percent write-down might prove disastrous for the health of certain institutions. Apparently, one of the major causes of the falling market is gossip about the growing probability of an accident somewhere in the financial system.

Prices and Income Losses and a Digression on Bank Capital (that is, Bank Equity)

Suppose a country is importing oil, and its price rises. A well-known approximation of the income loss of this country is:

$$\Delta Y = -M\Delta p_m$$

Here M is the amount of oil imported, and Δp_m is the change in the price of oil. Dividing both sides of this expression by Y, and multiplying the righthand side by p_m/p_m yields:

$$\frac{\Delta Y}{Y} = -\frac{p_m M}{Y}\left(\frac{\Delta p_m}{p_m}\right)$$

This is the transfer, in percentage terms, that is brought about by a change in the price of oil: if, for example, imports of oil are 5 percent of the income of a country and import prices rise by 100 percent, real incomes in the country importing oil will fall by 5 percent. Note, however, that if substitution away from oil takes place, then by the

Hicks-Allen term of the Slutsky expression, this loss will be decreased. The loss will also be smaller if the price of the oil-importing country's exports rise. To see this, consider the income of the oil-importing country to be expressed in units of its export good (such as machinery). Then we have:

$$\Delta Y = -M\Delta\left(\frac{p_m}{p_x}\right)$$

or:

$$\Delta Y = -M\frac{p_m}{p_x}\left[\frac{\Delta p_m}{p_m} - \frac{\Delta p_x}{p_x}\right]$$

As is clear from this expression, a high rate of increase in the price of the oil importing country's exports will, all else being equal, mitigate the loss of income of this country by reducing the purchasing power of oil—as mentioned earlier.

The importance of bank equity also requires some special attention. Consider the following balance sheet for a typical commercial bank, where the numerical entries are in dollars:

			(1)		*(2)*	
(A) Assets	*Liabilities (L)*		*A*	*L*	*A*	*L*
(R) Reserves	Deposits (D)		(R) 100	700 (D)	(R) 100	700 (D)
(C) Cash		=	(C) 50		(C) 50	
(S) Securities	Equity (E)		(S) 200	300 (E)	(S) 200	100 (E)
(L) Loans			(L) 650		(L) 450	
			1,000	1,000	800	800

The original balance sheet of the bank is shown in column 1. Suppose now that $200 in loans passed into the nonperforming category. This is bad news, as the write-down of this institution's assets (and equity) shown in balance sheet 2 makes clear. The owners of the bank (that is, the stockholders) have their equity reduced from $300 to $100, but the depositors' money is not in danger and, all else being equal, there is no reason for them to lose faith in the bank.

But what would have happened had $400 in loans passed into the nonperforming category. Once again a write-down will have to take place, but the question is what will happen with only $600 in assets (that is, reserves plus good loans plus salable securities plus cash) if all the depositors show up in front of the bank asking for their money, which is

$700. The first depositors reaching the bank can be paid, while those coming last will not get anything. In fact, the situation is more serious than it looks, because if the run on the bank is an impulse phenomenon, the bank will not have time to call in its loans. This situation is shown in balance sheet 3, and as is clear, equity is now negative.

(3)

A	*L*
(R) 100	700 (D)
(C) 50	
(S) 200	−100 (E)
(L) 250	

(4)

A	*L*
(R) 100	700 (D)
(C) 250	200 (L_b)
(S) 200	−100 (E)
(L) 250	

(5)
Central Bank

A	*L*
(L_b) +200	+200 (CASH)

A bank in this situation might therefore attempt to borrow from other banks. If this took place we would have the situation in balance sheet 4, where the assumption is that the loan was made in cash. Equity is still negative, but all depositors who come to claim their money can get it. When they find this out, many of them will put their money back in the bank, and the bank might eventually be able to rebuild its equity. The liability corresponding to the loan is called L_b, and I assume that the value of the loan is $200.

A problem that might arise here, however, is that there might be a run on all banks—in other words, a bank panic. In this situation, or even if this possibility exists, most banks will not be in the mood to extend loans to each other. It is possible though that loans can be obtained from the central bank. In fact, in order to dampen the panic, the central bank might declare that it is prepared to act as the lender of last resort, and given its money-creating powers, this is a guarantee that everyone asking for their money will be able to get it. The central bank book-keeping for this transaction is shown in balance sheet 5.

All this certainly looks simple, and in the bookkeeping sense it is; but what we have is, to a considerable extent, unwarranted money creation of the most flagrant kind that, according to some observers, could have dire inflationary consequences. Needless to say, this is not quite true, because the extra reserves provided at one point in the financial system can be drained away from the rest of the system. Whether this will happen, however, is quite another matter.

Appendix 1A
Production, Reserves, and the Reserve/Production Ratio

Table 1A-1 shows production, reserves, and the reserve/production (R/P) ratio for the largest oil-producing countries. The R/P ratio is thoroughly discussed later in this appendix.

OPEC oil production has perhaps fallen to 16 mbbl/d at various times during the past few years. The country in which output has fallen the most rapidly, although a slight recovery has been noted, is Iran; but production has also fallen rapidly in Saudi Arabia because of its role as OPEC's swing producer. On the other hand, in the non-OPEC countries, production has been steadily increasing, and therefore OPEC's share of the world market has been substantially reduced over the past few years. Production has risen especially fast in Western Europe, where the 1983 increase over 1982 came to approximately 11.4 percent. Everyone, however, does not believe that this kind of increase is sustainable. The most recent issue of "The World Energy Outlook Through 2000," published by the Continental Oil Company (CONOCO), maintains that non-OPEC oil output will peak in the late 1980s and then decline slowly. By the same token, OPEC production will grow to almost 29 Mbbl/d in the year 2000. CONOCO also believes that as OPEC production approaches its technical or politically determined maximum, the oil-importing countries will become increasingly vulnerable to supply disruptions and sudden price increases. CONOCO also appears to believe that by 1990 the United States will be importing substantially more oil than it is today.

U.S. production is now static, which if nothing else reflects the advanced stage of U.S. oil fields in their depletion cycle. There have also been several resounding exploration failures in the United States, including a $1 billion dry hole at Mukluk in Alaska's Beaufort Sea. Some people interpret the Mukluk disappointment as the start of a definitive oil production decline in the United States.

Crude output is fairly high in Venezuela, where it appears that the OPEC production quota is not always respected. On the other hand, Brazil is unaffected by such things as quotas, and is aiming at oil self-

Table 1A–1
Production, Reserves, and Reserve/Production Ratio of the Largest Oil-Producing Countries

Country	*Production (1983) (kbbl/day)*	*Reserves (1/1/1984) Gbbl*	*R/P Ratio (Years)*
Canada	1,585	6.7	13
United States	10,248	27.3	9
Total North America	11,833	34.0	9
Argentina	480	2.4	14
Brazil	337	1.8	16
Mexico	2,953	48.0	49
Venezuela	1,801	24.9	38
Total Latin America	6,393	81.7	37
Algeria	684	9.2	37
Egypt	724	3.5	14
Libya	1,075	21.3	57
Nigeria	1,236	16.6	37
Total Africa	4,404	56.9	34
Abu Dhabi	—	30.4	110
Iran	2,409	51.0	54
Iraq	934	43.0	130
Kuwait	1,094	63.9	192
Neutral Zone	—	5.7	39
Oman	377	2.8	20
Qatar	—	3.3	33
Saudi Arabia	5,074	166	93
Total Middle East	11,524	370.1	87
Norway	625	7.7	35
United Kingdom	2,347	13.2	16
Total Western Europe	3,403	23.0	20
Australia	417	1.6	11
Brunei	—	1.4	25
India	480	3.5	25
Indonesia	1,301	9.1	19
Malaysia	365	3.0	22
Total Asia and Australia	2,783		
USSR	12,485	63.0	14
China	2,123	19.1	25
Total centrally planned economies	14,988	84.6	16
Total: Noncentrally Planned Economies	40,340	584.7	42
Total World	55,328	669.3	34

Source: *The Petroleum Economist* (various issues): *The OPEC Bulletin*, 1983, 1984; *British Petroleum Statistical Review of World Energy*, 1974.

Note: totals include statistics for smaller producing countries. Production in thousands of barrels per day. For example 10,248 thousand = 10.248 million.

sufficiency in ten years. Production is static in Argentina, where the main energy hope is natural gas.

Production is rising in a number of the smaller African countries: Angola, Cameroons, the Ivory Coast, and perhaps Zaire, while the production of Gabon appears to be stable. The output of Egypt is increasing, but the major strike that the Egyptians have desired for so long, and also expected, has still not taken place, although there is a great deal of prospecting in that country. In the Far East, Australian production is rising; but it is Malaysia where output is increasing at the most rapid rate in the region. There are persistent rumors of a major oil strike that could take place off the Malaysian coast at any time. For years so-called insiders have claimed that there is a huge natural gas bubble somewhere inside or in the vicinity of Malaysia that is just waiting for some lucky prospector.

China is now reaping the harvest of Western offshore experience, thanks to the Western firms that have been engaged to prospect in coastal waters off the People's Republic. As noted at some length in my book, *The Political Economy of Oil*, the pessimistic reports of the Central Intelligence Agency (CIA) concerning the amount of oil that the Russians will be able to produce in the near future did not make a great deal of sense, and apparently even the CIA has had some second thoughts. It is not unthinkable, however, that the Russians may experience some difficulty in meeting their production goals in the not too distant future, and this—plus the desire to maintain their exports of crude oil and oil products at their present level as long as possible—might explain the intense effort they are putting into substituting nuclear energy and natural gas for crude oil. There is also a great deal of energy to be won in conservation in the USSR, since energy intensities in that country are so high, relative to energy intensities in the same activities in the West, as to suggest that the Soviets are using many inefficient techniques in their heavy manufacturing industries.

In the long run, the important factor on the world oil scene will be reserves and not production. As reserves contract, and a larger percentage of them are concentrated in the Middle East, a new OPEC may come into existence. If the demand for oil eventually develops as many of us think it will, this new OPEC will be able to exercise a stronger control over the oil price than was imaginable by the present OPEC.

We can now employ some elementary algebra and differential calculus to clarify an extremely important aspect of the exploitation of a single country's oil deposits. In the simplest possible language, what it comes down to is as follows.

For technical reasons (which I show in *The Political Economy of Oil* to have a direct economic relevance), the output per period from a given deposit of oil should not be larger than a fraction of the total stock of oil in the deposit. Experienced oil producers say that this is usually about one-tenth, although it can be slightly larger or smaller, depending on the geological characteristics of the deposit. For instance ARAMCO (the Arabian–American Oil Company) customarily used one-tenth for Saudi Arabia.

If, for example, we have a field containing 150 units of oil to begin with, and we desire to lift 10 units a year, we can do so without violating the one-tenth criterion during the first five years. In these years reserves fall from 150 to 100 units, and the reserve-production (R/P) ratio, measured at the end of the year, falls from 14 to 10, as reserves fall by 10 units a year. (That is, the reserve production ratio is 140/10, 130/10, ..., 100/10.) But after the fifth year, if we were to continue producing 10 units per year, the R/P ratio would become less than 10. (Put another way, we would be lifting more than one-tenth of the field's reserves in a single year.) As a result, from the beginning of the sixth year, the critical R/P ratio takes over and determines production in that period and all ensuing periods.

To see how this goes, if production were 10 units in the sixth period, reserves would fall to 90 units at the end of that period, and the R/P ratio—calculated at the end of the period—is 9. To get a production for an arbitrary period t that will prevent the R/P ratio from falling under a certain value θ^* (the critical R/P ratio), it is necessary that:

$$\frac{\text{End of period reserves } (t)}{\text{Production } (t)} = \frac{\text{End of period reserves } (t-1) - \text{Production } (t)}{\text{Production } (t)} \geq \theta^*$$

Taking R as reserves, and thus R_{t-1} as reserves at the end of period $t - 1$, and P as production, and thus P_t as production in period t, this expression can be simplified immediately to give P_t. Remember, however, that this relationship is not used during those periods when the R/P ratio is greater than or equal to the critical R/P ratio θ^*. Accordingly, in the numerical example presented here, it is not used in the first five periods.

$$P_t \leq \frac{R_{t-1}}{(\theta^* + 1)} \tag{1A.1}$$

Continuing with the same example, where we need a production for the sixth period that will not cause the R/P ratio to fall below $\theta^* = 10$, we use the equality sign in equation 1A.1 to obtain $P_6 = R_5/11 = 100/11 = 9.09$. End-of-period reserves in year 6 are then $100 - 9.09 = 90.01$. Accordingly, production in the seventh year P_7 must be equal to $R_6/11 = 90.01/11 = 8.35$. Figure 1A-1 and its accompanying table show what happens to production, reserves, and the R/P ratio.

Equation 1A.1 is a refinement of one constructed in *The Political Economy of Oil*. Its purpose here is to provide some background for my main point. That is, in a country with a number of oil fields of different sizes, it can happen that if the total oil production of the country turns down because output must be decreased in a large field that has reached its critical R/P ratio, the R/P ratio for the country's entire stock of oil when this decline takes place can be larger than the critical R/P ratio for a single deposit, and perhaps much larger. For example, in discussions in the mid-1970s of the ultimate annual production of crude oil in Saudi Arabia, it was generally overlooked (outside of Saudi Arabia) that for the Saudi Arabian oil sector as a whole, output might start to fall when R/P was in the vicinity of 15, even though the critical R/P ratio for individual deposits was about 10. This phenomenon cannot be completely explained here, but an important factor was the presence of perhaps the world's largest oil deposit, Ghawar, together with a number of smaller fields where—for economic reasons—it may not make sense to undertake investments to maintain total production once the output of Ghawar starts to decrease. Also, in equation 1A.1, the life of the deposit is not $150/10 = 15$ years, despite the fact that this kind of reasoning is constantly used in the media. In this and similar cases, the life of the

Year	*Production*	*Reserves*[a]	R/P
1	10	140	14
2	10	130	13
3	10	120	12
4	10	110	11
5[b]	10	100	10
6	9.09[c]	91.9	10
7	8.35	83.5	10
8	7.59	75.9	10
9	6.90	69.0	10

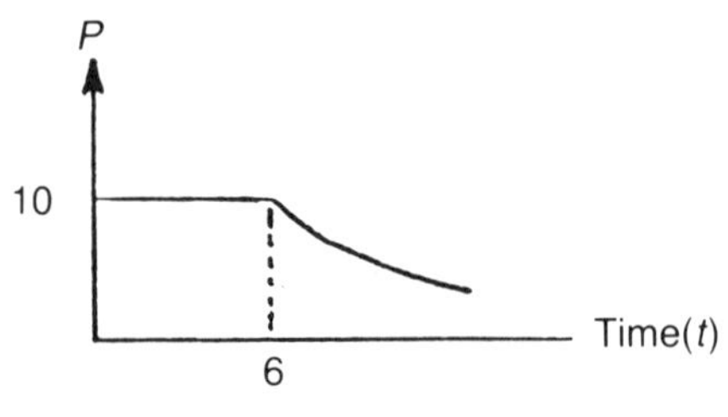

[a] End of year figures.
[b] Critical R/P ratio reached at end of year 5.
[c] Production starts to fall in the beginning of period 6 (or at the end of period 5).

Figure 1A–1. Production Profile with a Critical R/P Ratio of 10 (θ^* 10)

deposit is infinity. However, an even more important observation is that in the sixth year, when production turned down, there were 90 units of reserves in the ground out of an original 150.

There are several ways to discuss the R/P ratio for the entire oil-producing sector of a country, but to bring out the basic issues, I will present two rather simple examples that can be turned into numerical exercises like equation A.1. First, take a country with a single large field and a number of smaller fields that contain the same amount of reserves and have the same output. Then assume that at the time the critical R/P ratio ($\theta^* = R'/P'$) is reached for the large field, the R/P ratio for each of the other fields (R''/P'') is greater than θ^*. Thus, for the country as a whole, we have the following expression for the R/P ratio:

$$\bar{\theta} = \frac{R' + \sum R''_i}{P' + \sum P''_i} = \frac{R' + nR''}{P' + nP''}$$

The assumption here is that there are n small deposits of the same size. Now note that if $n \rightarrow \infty$, then $\bar{\theta} \rightarrow R''/P''$, and this is assumed to be greater than θ^*. Also, if $n = 0$, then $\bar{\theta} = R'/P' = \theta^*$. Now differentiate this expression with respect to n (to find out what happens when n increases):

$$\frac{d\bar{\theta}}{dn} = \frac{P'R'' - P''R'}{(P' + nP'')^2}$$

If $P'R'' - P''R' > 0$, then $d\bar{\theta}/dn > 0$. But $P'R'' - P''R' > 0$ means $R''/P'' > R'/P'$, and this is the assumption with which we began the demonstration. We therefore see that we must have $\theta^* < \bar{\theta} < (R''/P'')$ at the time when the critical R/P ratio is reached for the large deposit. Obviously, $\bar{\theta}$ could be fairly large when production started falling.

Finally, it might be useful to approach this problem from a slightly different point of view. Suppose that at time t_o we have an initial amount of reserves R_1 that are being depleted at a rate of P_1 units/period. Now assume that after t' periods we arrive at the critical R/P ratio θ^*: that is, at time $t = t'$ we have $\theta_1 = \theta^*$, where θ_1 is the R/P ratio for this deposit at any given time. Thus:

$$\frac{R_1 - t'P_1}{P_1} = \theta_1 = \theta_1^*$$

At the same time, suppose that another deposit that at time t_o contains R_2 units of reserves, and the production from this deposit is P_2 units per period. At time t' the R/P ratio of this deposit is θ'_2, and the assumption is that $\theta'_2 > \theta^*_1$. By definition the R/P ratio of this deposit is:

$$\frac{R_2 - t'P_2}{P_2} = \theta'_2$$

The aggregate R/P ratio $\bar{\theta}$ at time t' is:

$$\bar{\theta} = \frac{(R_1 + R_2) - t'(P_1 + P_2)}{P_1 + P_2} = \frac{(R_1 - t'P_1) + (R_2 - t'P_2)}{\mathrm{P}_1 + \mathrm{P}_2}$$

This expression, together with the above relationships, results in:

$$\bar{\theta} = \frac{\theta^*_1 P_1}{P_1 + P_2} + \frac{\theta'_2 P_2}{P_1 + P_2} = \beta_1\theta^*_1 + \beta_2\theta'_2$$

Here, $\beta_i = P_i/(P_1 + P_2)$, with $i = 1{,}2$ and $\beta_1 + \beta_2 = 1$. Given the conditions of our exercise ($\theta'_2 > \theta^*_1$), this means that $\theta^*_1 < \bar{\theta} < \theta'_2$. Thus we have a situation where the aggregate R/P ratio $\bar{\theta}$ is larger than the critical R/P ratio of the first deposit, when total production turns down due to a turning down of the production of the first deposit. In general, with n deposits, we have:

$$\bar{\theta} = \sum_{i=1}^{n} \beta_i\theta'_i \qquad (\text{with } \theta'_1 = \theta^*_1)$$

In the first example, it was noted that when production turned down, three-fifths of the original reserves were still in the ground. This fraction could be larger in a multideposit situation, something that should be remembered if oil consumption should begin to increase and R/P ratios start to fall rapidly again.

2
An Introduction to Coal

Coal is formed from the remains of trees that have been preserved for millions of years under a special oxidizing condition, for the most part in swamps where, after falling, the trees either did not rot or rotted very slowly. The conversion process itself took place by photosynthesis. In general, coal seams lie on so-called underclays that are associated with coastal swamps; and their quality is largely, but not entirely, a function of their age. Top-grade coal requires a gestation period of a few hundred million years, and it has been calculated that the average time required to accumulate enough vegetable matter to form 1 meter of coal is about 1.6 million years. Similarly, a coal seam 1 meter thick would have been compacted originally from a 120-meter layer of plant remains.

It is possible to distinguish a spectrum of coals, ranging from peat through anthracite. *Peat,* which is brown, porous, and often contains visible plant remains, is the lowest class of coal, with an average energy content of 4,000 Btu/lb. (Btu signifies the heating unit British thermal units. Energy units will be taken up directly below and in appendix B.) Peat also has a very high moisture content. *Lignite* has a moisture content that averages 41 percent and an average heat content of 7,000 Btu/pound. *Bituminous* coals, on the other hand, are characterized by a low moisture content, and the moisture content of a typical anthracite coal is only 3 percent. Where energy values are concerned, we distinguish between *subbituminous* coal, with an average energy value of 9,000 Btu/pound, and *bituminous* coal proper, with an average energy value of 12,000 Btu/pound. *Anthracite* coal, which is jet black and difficult to ignite, has an average energy value of 13,720 Btu/pound.

The World Energy Conference has subdivided coal into two major calorific categories: *hard* coal, defined as any coal with a heating value above 10,250 Btu/pound on a moisture and ash-free basis; and *brown* coal, which is any coal with a lower heating value. This puts subbituminous coal into the brown coal category, which is generally unacceptable in the United States and Canada. Still another system divides coal into the classifications *brown coal* and *black coal.* Brown coal is geologically young, high in water content, and comparatively speaking has been subjected to little pressure; while black coal has been subjected to more intense pressure, is considerably lower in water content, and contains more carbon. Black coal ranges from subbituminous coal (which is usually dull black and waxy in appearance) to anthracite, and is divided into two general categories: coking or metallurgical coal, and thermal or steam coal. East Germany is the world's largest producer of brown coal, with 250 million tons/year, followed by the USSR with 180 million tons/year. Among the non-Communist countries, West Germany is the largest producer of brown coal and lignite; but Australia's resources of brown coal are probably much larger than either of these two countries. The thickest seam of brown coal in the world is at Loy Yang in Australia.

Several units are used to measure energy. Physicists prefer joules, while engineers are partial to British thermal units (Btu) or kilowatt hours (kwh). The transformation between joules and Btu has been carefully measured: 1 Btu $= 1.055 \times 10^3$ joules. Since different coals have different calorific contents, as made clear previously, a standard measure of coal and energy use has been defined. This is the *ton of coal equivalent* (tce), which is defined as a metric ton (1 metric ton = 1 tonne = 2,205 pounds) of coal with a specific heating value of 12,600 Btu/pound. Consequently, more than one metric tonne of coal might be necessary to produce the heating value of 1 tce. For example, 1 tce = 1.4 tonnes of subbituminous coal, using the heating value of 9,000 Btu/pound given earlier. Consider also that in 1977 world coal production came to 3,400 million metric tons of raw coal, which was 2,500 million metric tons of coal equivalent (2,500 Mtce), or 33 million barrels of oil per day (33 Mbbl/d). This last figure is obtained from the following equivalency between oil and coal: 1 tce converts to 4.8 barrels of oil, and 76 Mtce/year is equivalent to 1 Mbbl/d of oil. To a certain extent, tce is an artificial unit, since its heating value is probably higher than the heating value of an average tonne of coal extracted during any given year. In addition to its energy content, coal is graded with respect to the amount of waste materials it contains—primarily sulfur and ash—and also its gas content. Table 2–1 lists some characteristics of coals with respect to these criteria.

Coal is principally used in the generation of electricity and the production of heat (through direct burning), in metallurgical processes, and as a feedstock in the production of coal gas and synthetic oils and fuels. In the United States in 1972, 57 percent of all coal mined was burned in electricity-generating plants; 12 percent was used as coking coal, and went to the making of coke (a strong, porous block) in the coking ovens of the steel industry; and 10 percent was exported. A very small amount was used for other purposes, such as the making of synthetic dyes. The production of synthetic oil and gas from coal takes

Table 2–1
Some Physical Characteristics of Coal

	Carbon Content[1] *(Dry Ash-Free)*	*Volatile Matter*[1] *(Dry Ash-Free)*	*Calorific Value*[2] *(Ash-Free)*	*Moisture Content*[1,3]
Peat	60	53	16,800	75
Brown coal or lignite	60–61	53–49	23,000	35
Subbituminous coal	71–77	49–42	29,300	25–10
Bituminous coal	77–87	42–29	36,250	8
Anthracite	81–91+	29–8	36,250	8

Source: The Australian Coal Association.
[1]Average percentage.
[2]Kilojoules/kilogram.
[3]In situ.

place only on a pilot scale in most of the industrial world; but the Republic of South Africa already has a commercial installation for this purpose operating at Sasolburg, and the El Paso Power and Light Company has leased the South African process with the intention of perfecting it and producing synthetic gas and liquids in the United States.

On a worldwide basis, the major portion of coal production is of hard coals (70 percent), with most of this output earmarked for thermal purposes. A small and declining proportion of hard coal (about 25 percent) is of the metallurgical variety. The best coking coal is bituminous coal with a low or medium gas content and a low sulfur content, since large amounts of gas and sulfur contaminate the charge of ore and limestone that coking coal is supposed to support. Almost all brown coal is used for power generation, but since in general this type of coal cannot be stored or transported because of the danger of spontaneous combustion, it is usually fed into power stations very close to the mining site. Another possibility is to turn it into briquettes by compressing the coal and shaping it into blocks. In this form, brown coal can be transported without being constantly doused with water, and its energy content is also increased with reference to its weight: typically, brown coal from the Latrobe Valley in Australia has its energy content increased by a factor of between two and three when it is briquetted.

Because of the expense or lack of facilities for transporting coal, the tendency of late has been to locate as many coal-based power stations as possible close to coal fields. Transport expenses also explain why most of the nonmetallurgical coal entering into international trade is anthracite, since shipping coal with a low energy content is patently unprofitable. Still, a huge amount of coal must be moved over large distances every day—for the most part by train, but also in pipelines (as a slurry); and one of the reasons for the relatively slow expansion of the world coal industry is the difficulty in financing the investments in coal-transporation facilities that would be required if coal consumption were expanded by the amount proposed by the International Energy Agency (IEA) and similar organizations. This problem is aggravated in very large countries like the United States where the coal mining districts are well inland, and to a much greater extent in the USSR.

There are three major methods for mining coal. The first is *underground mining*, whereby a vertical or inclined shaft from the surface intersects a coal bed of satisfactory thickness and quality. Miners and their equipment enter this shaft, drive lateral openings from it into the coal face, and extract the coal. Exploitability, which depends on such things as seam thickness, depth, and type of terrain, varies in underground mines from 25 to 70 percent. Strip mines, on the other hand, are developed by removing the overburden covering a minable deposit of coal, then breaking up the deposit by digging or blasting, and finally removing the broken coal. In general, strip mining (which is sometimes called *open-pit* or *open-cast* mining) is much less costly per ton of coal produced than underground mining, even when the land disturbed by stripping is completely reclaimed.

The third method, which is applicable to only a few coal deposits, is *auger mining*. Large horizontal augers bore into coal beds that are exposed on the side of hills and push out broken coal as they rotate. Obviously, this kind of technique can only be used in special situations. At present in the United States, more than half the production of coal results from strip mining; and this rate will increase as the center of gravity of U.S. coal mining shifts westward toward areas where strip mining is the appropriate method of exploitation. An interesting and important feature of strip mining is its capital intensity. The expansion of the Australian coal-mining sector has been accompanied by a large scale shift to open-cast mining; as a result, there has been a significant increase in the productivity of labor as the amount of capital per unit of labor has gone up, as well as a reduction in the number of employees on the newer mining sites.

By way of concluding this discussion, a few details will be given about three coal-producing operations: the Selby Coal Field in the Yorkshire district of England, the Jang Seong colliery in Korea, and the Belchatow coal field in Poland. The first two of these are underground mining complexes, while the third is open pit. By the standards of the countries in which they are located they are extremely large and intended to greatly enlarge domestic coal production.

Britain still possesses impressive reserves of coal, and in certain quarters it has been claimed that there is enough coal in that country to satisfy total energy needs through the twenty-first century. But British coal is high-cost coal and essentially noncompetitive in relation to—for example—South African coal. One of the purposes of the Selby scheme is to show that coal can be produced in the United Kingdom at acceptable costs. The Selby coal field comprises 110 square miles of high quality bituminous coal, and at peak production a work force of 4,000 is supposed to extract 10 million tonnes (10 Mt) of coal a year, which is one-twelth the current output of coal that is produced under the jurisdiction of the National Coal Board (NCB). The output per manshift in the Selby fields is expected to be five times the current United Kingdom average, and in theory it will only take 90 minutes for a piece of coal cut from the pit to reach the nearby Ferrybridge power station. This station is one of three that will receive coal from the Selby development, with the other two being Eggborough and Drax; Drax is considered one of the most modern power stations in the world.

Although Yorkshire was thoroughly explored in the first part of this century, the existence of the Selby field was overlooked until the 1960s. At that time, surveys revealed a field containing 2 billion tons of coal at workable depths. The seam itself is known as the Barnsley seam, and its depth varies from 250 meters in the west to 1,100 meters in the northeast. The thickness of the seams is between 2 and 3.5 meters, and coal from this seam is clean and contains very little waste rock. The basic piece of equipment for each mininig team is a *shearer loader,* which cuts coal like a bacon slicer. The five pits comprising the Selby complex can produce a total of 2,000 tonnes a day, and all operations are highly automated. Among other things this means that miners working at a coal face are enclosed in a protective cocoon of steel, and mine disaster scenes of the type depicted in such classics as *How Green Was My Valley* are generally considered unthinkable.

On the other hand, deaths and injuries are far from unknown at the Jang Seong pit in Korea (which is operated by the state-owned Dai Han Coal Corporation). This mine, the largest in South Korea, is a key element in the plans of the Korean government to promote coal use in the household sector and convert several very large oil-fired power plants to coal. Present intentions call for a doubling of the proportion of coal-based power generation, in step with a halving of the proportion of oil-based power generation. Obviously, nuclear power will also play a large part in the South Korean power picture.

At present, the main operation taking place at this mine is the sinking of a 3,000-foot shaft at the rate of 120 feet/month. This is part of a scheme to keep the output of the mine from suddenly falling about the end of this decade; however, if present plans to expand the use of coal are fulfilled, considerable imports of coal will be required regardless of what happens at the Jang Seong or elsewhere in the domestic coal sector. Most of the coal mined in South Korea is anthracite, which is turned into briquettes for household use, and although two 200-megawatt (200 MW) anthracite coal firing plants are to be part of South Korea's coal-based electricity generating facilities, the main units will be 2,000-MW plants using bituminous coal imported from Australia, South Africa, and perhaps Canada.

Recent political events in Poland have slowed down developments at Belchatow, but some authorities still claim that the Belchatow operation will produce as much brown coal each year in the last part of the 1980s as the total brown coal production of Poland in 1979. Furthermore, this coal is to be fed directly into Europe's largest coal-burning power plant, which is expected to have a capacity of 4,350 MW. Within twenty years intentions are that brown coal will provide 40 percent of Poland's power, thereby freeing large quantities of high-quality hard coal for export. The Polish mining industry lacks the sophistication of—for example, that of the English—and the environmental problems posed by coal tend to be ignored. The main technical problem faced by the coal industry is provided by the transport network, which is heavily overloaded and badly in need of repairs.

Reserves, Resources, and Production

Reserves are essentially the known inventory of a natural resource that can be economically extracted. This definition is often extended to include unknown materials that, when discovered, will be no more expensive to exploit than those currently being won (such as any oil that will be found in Saudi Arabia in the near future), and known materials that are slightly submarginal economically but will probably be promoted to the category of exploitable reserves as a result of technological progress or a rise in their price. Everything else can be labeled resources. The U.S. Geological Survey (USGS) has compiled estimates of fossil-fuel resources on the basis of such things as the anticipated frequency of these resources in certain types of rock formations. Resources also include known accumulations of a material whose grade is too low to warrant exploitation at any time in the forseeable future, but which might be exploitable at some distant point in the future because of the radically different application of a known technology.

The formalization of these concepts is due to Saul G. Lasky, who first made the important distinction between reserves and resources in

1949. Hitherto these terms had been used interchangeably. In 1956 Blondel and Lasky offered the following relationship:

$$\text{Resources} = \text{Reserves} + \text{marginal resources} + \text{submarginal resources} + \text{latent resources}$$

A slightly more complex breakdown can be found in figure 2–1, which was developed by the U.S. Geological Survey. On the horizontal axis are total supplies, or resources as they are defined in the Blondel-Lasky relationship; while on the vertical axis the characteristic of interest is economic value. It therefore seems logical that where this characteristic is concerned, the natural partition is between profitable and unprofitable.

The World Energy Conference has also provided a set of definitions that are analogous to, but not the same as, those described previously:

A. Total resources: "The total quantities available in the earth that may be successfully exploited and used by man within the foreseeable future." For Britain, in 1977, coal resources were put equal to 190×10^9 tonnes.
B. Total known reserves in place: "The corresponding fraction of resources that have been carefully measured and assessed as being exploitable in a particular nation or region under present local economic conditions using available technology." Total known reserves in place were estimated to be 100×10^9 tonnes for Britain in 1977.
C. Recoverable reserves: "That fraction of reserves in place that can be recovered under the above economic and technical limits." Recoverable reserves of coal were put at 45×10^9 tonnes in 1977 for Britain.
D. Additional resource: "All other classifications with a lower degree of geologic certainty as to their existence than those indicated as known." These were also estimated to be 45×10^9 tonnes.

		Calculable		Estimated	Hypothetical	Speculative
		Known	Probable			
Profitable		Reserves				
Unprofitable	Para-marginal					
	Sub-marginal					

Figure 2–1. The Classification of Mineral Supplies

Table 2–2 includes estimates of reserves, resources, and production. Among other things, it shows that over 60 percent of reserves are concentrated in just four countries, and the same is true of 90 percent of known resources—although coal deposits can be found in more than eighty countries. Resources are concentrated in the USSR (45 percent), the United States (24 percent), China (13 percent), and Australia (6 percent). Of the less developed countries, India is perhaps the only one with a major coal-producing and exporting capacity, although some geologists believe that Colombia, Indonesia, and Botswana will eventually export a considerable amount of coal. As things now stand, only a few countries will be very high up in the coal-exporting league before the world is well into the 1990s. Three of these will, of course, be the United States, Australia, and South Africa; and they will be joined by Poland and perhaps Canada. Rumor also has it that Colombia will be a large exporter by the year 2000.

Of the countries named in table 2–2, the United States must rank as the number one coal power, regardless of the patently larger amount of resources possessed by the USSR. In the United States there are huge reserves located in a favorable climate, and proximate to a highly efficient transportation network. In addition, with the gradual shift in coal-mining operations from underground installations in the East to open-pit mining in the West, U.S. unit mining costs could easily fall. This is because labor productivity in open-cast mining is often several times that of underground mining. (Still, it should be recognized that at present 70

Table 2–2
Reserves, Resources, Production, and Exports of Coal

Country	*Reserves*[1]	*Resources*[1]	*Production*[2]		*Exports*[2]	
			1982	*1983*	*1982*	*1983*
United States	167	2,570	752	703	96	70
China	99	1,438	655	692	3	7
Australia	33	600	92	102	48	57
Canada	4	323	43	43	16	17
West Germany	34	247	88	82	10	—
United Kingdom	45	190	125	119	7	7
Poland	60	140	189	191	29	35
Spain			38	—	—	—
Republic of South Africa	43	72	140	156	27	25
USSR	110	4,860	710	718	22	—
India	12	81	—	—	—	—

Source: Australian Department of Trade and Resources, Annual Reports, 1982, 1983; United States National Coal Association.

[1]In Gtce for 1982.

[2]1983 estimated; production and exports in Mtce.

percent of U.S. reserves can be reached only by underground mining.) Zimmerman (1982) has also shown that the effect on costs of depletion of reserves west of the Mississippi is very small: if 120 Mt/year is produced for twenty years in the Power River Basin, coal production costs are only increased by 8 percent over 1973 costs; and similar results were obtained for other districts. On the other hand, in certain parts of Appalachia, as more low-sulfur coal is produced, costs can be expected to rise significantly. According to Zimmerman, the cost of low sulfur coal in much of central and southern Appalachia would increase by 51 percent if cumulative production were increased by a factor of 20.

According to the Energy Modeling Forum at Stanford University, by the year 2000 the share of the states west of the Mississippi River in U.S. coal output should be up to at least 60 percent, compared with a current value of less than 35 percent. Looking exclusively at supply constraints, coal production in the United States is capable of expanding by a factor of 3 or 4 by the beginning of the next century; but doubts are now been expressed as to whether there will be a demand for this much coal. The demand for coal is heavily dependent on the requirements of power stations, and at least in the near future the deceleration in the world economy will reflect on the amount of electricity required. Of course, the sharp increase in the price of oil has stopped the substitution of oil for coal that was taking place in all countries with a substantial electricity-generating capacity; but the conversion of oil-fueled installations to coal has proceeded much more slowly than originally expected, and the recent softening of the oil price has worked to make many power station managers completely lose interest in the highly touted attractions of coal. But even so, much oil-fired electricity generating capacity will undoubtedly be replaced with coal-fired installations.

As for the USSR, coal production during the 1970s increased by about 2 percent a year. It is expected that great efforts will be made to raise this figure, but some doubts exist as to whether the Soviet transport system and the competence of the Soviet bureaucracy are equal to the task. At present eastern Europe and Japan (the world's largest importer of coal) purchase fairly large amounts of Soviet coal, and predictions are that Japan will eventually be interested in a much larger share of the Soviet output. With this customer, however, the USSR faces increasing competition from both China and Australia, and possibly from Canada, since Japanese economic policy is distinguished by a consistent and conscientious effort to diversify the sources from which it buys coal.

The Australian situation will be taken up in detail in later chapters. Because of its political reliability, stability, and relatively small needs, Australia is a major potential source of energy materials for the entire industrial world. China, in contrast, has enormous resources; but no one knows just when or how these will be developed or whether they will be put on the export markets. The largest coal fields in China are in the north of the country, although other regions probably contain sizable deposits. Chinese sales of coal to Japan may also be increasing, but a major constraint on these dealings is the limited capacity of the Chinese rail system as well as the lack of facilities at ports for handling large amounts of coal.

Exploration and the Length of Life of Reserves

The recoverable reserves of coal can also be considered in terms of their length of life. On the basis of consumption in 1978, there was enough hard coal available to last about 200 years. Similarly, with constant consumption, there was enough lignite to last about 285 years and subbituminous coal to last 780 years. In considering this matter, two interesting and important questions are raised. The first has to do with exploration, while the second concerns the dynamic length of life of coal reserves—or how long a given stock of coal will last if its consumption is growing by a certain percentage every year. As will become apparent, with positive rates of growth of consumption, the figures presented earlier in this paragraph are not very useful.

In considering the first of these issues, we see immediately that on the basis of simple textbook criteria of social (as compared to private) profitability, it does not make sense to search for reserves that will not be consumed until scores, or hundreds, of years in the future. Remember that a dollar invested in searching for coal could also be invested in a government bond or some other safe financial or physical asset (or for that matter it could be used to increase present consumption). Thus, even if it were certain that the dollar spent for exploration would result in the availability of a new unit of coal, that coal would have to sell for an amount which, after deducting the capital and labor costs required to remove the coal, resulted in a larger revenue for the exploration department of a firm than would have been obtainable had the money used for exploration been invested in something else (such as a safe financial asset). But even if this criterion is met, it may still be socially unprofitable for a firm with a thirty-year supply of coal to invest in finding more coal, since perhaps society can obtain its coal requirements in thirty years from the reserves of other firms and obtain them at a lower price.

But in the real world, plagued as it is by uncertainty and imperfect markets, political or strategic considerations can nullify this kind of logic. Despite the amount of coal already discovered globally, individual countries (or private firms) that desire to consume or sell a great deal of coal in an uncertain future find it perfectly sensible to invest in exploration as long as they feel that the general energy situation is such that they might not be able, at some point in time, to obtain supplies that they cannot do without. Countries like Poland and East Germany are obviously in this predicament; and it is fairly easy to devise situations where the same is true for private enterprises.

For example, if a mineral is used as an input in some type of activity involving expensive durable capital goods, enough of the resource must be available to allow the utilization of that equipment over the greater part of its lifetime. Take, for instance, a country like Sweden that is considering replacing its nuclear power plants with coal-fired installations. This decision is rational by conventional economic standards only if it is believed that in the future enough coal will be available at reasonable prices to make the generation of electricity by coal-fired plants as inexpensive as nuclear-based electricity. (It could, of course, be rational on non-economic grounds). Clearly, employing equipment with a lifetime of thirty years for only fifteen or twenty years would generally (though not always) be a very unprofitable undertaking. Moreover, if coal were becoming scarce, or if it were plentiful but consumers believed it was becoming scarce, individual coal sellers might find themselves in a position where, in order to sell coal, they would have to guarantee deliveries many years in the future. Thus individual firms might invest heavily in exploration, although there was a sizable glut of the commodity on world markets.

The last matter to be considered in this section is the dynamic length of life of a mineral deposit. If, for example, there were 100 units (such as tons) of a deposit, and they were extracted at two units/year, then the deposit would last fifty years. But if consumption of the mineral grew by 3 percent a year, starting from two units, then the deposit of 100 units would only last thirty years; while if consumption grew by 10 percent a year, then the deposit would only last seventeen years. Thus it is apparent that sizable expected growth rates of consumption can greatly alter the length-of-life figures presented at the beginning of this section. Furthermore, a nonzero rate of growth in consumption causes even large additions in the amount of a deposit to have only moderate effects on the life of deposits. If, to continue this numerical example, the amount of the deposit were increased by a factor of ten (from 100 units to 1,000), with a growth rate of consumption of 10 percent the length of life of the deposit would only be increased by a factor of 2.3—that is, from 17 years to 39.3 years. Thus, even if all the coal in the world classified as resources could be reclassified as reserves, and in addition new supplies were discovered, coal is still a very finite resource, and loose talk about there being enough coal in the world to last "forever," "indefinitely," or for "hundreds of years" should be ignored. This situation is further emphasized in chapter 6, which includes an exact calculation for the length of life of coal reserves.

Discounting and Capital Values

Several topics in this book lend themselves to algebraic expositions, and in keeping with my announced intentions to separate technical and descriptive materials, the great majority of these expositions can be found in chapter appendixes. But the matter of discounting is basic to a great deal of energy economics, and in addition is well within the reach of readers with a minimal background in secondary school mathematics.

This approach to discounting need involve no more than formalization of the well-known fact that a given amount of an asset (including a sum of money) is worth more today than the same amount is worth in the future, all else remaining equal. (For instance, a given sum of money is not worth more today than it will be worth next year to me if I thought that the price level would fall by 50 percent.) The reason this is so is because if we have the asset now, we also have the option of enjoying its services if we so desire; otherwise we face uncertainties concerning our desires and appetites up to and including the date on which it is received. (Consider, for example, an automobile.) Usually, if people have a choice, they demand some sort of compensation for postponing present satisfaction. Thus a sum of money (which represents generalized purchasing power) today is related to a certain sum in the future through a discount or interest rate that says something about the actual compensation available for someone prepared to defer a unit of consumption for a given period. This period is generally taken to be a year. For instance, if I have \$100 today and the interest rate is 10 percent, then it is customary to say that this \$100 is equivalent to $100(1 + r) = 100(1 + 0.10) = \$110$ in a year's time. If I give up \$100 on June 13, 1984, I can obtain a bond or bank deposit that will provide me with \$110 on June 13, 1985. My compensation for waiting a year before spending my \$100 is therefore \$10.

It is also enlightening to turn this formulation around. That is, \$110 received a year from now is worth $110/(1 + r) = 110/(1 + 0.10) = \$100$ today. The general expression that we are moving toward is:

$$A_{t+1} = A_t(1 + r) \text{ or } A_t = \frac{A_{t+1}}{(1 + r)}$$

In this expression, t signifies the time period. Similarly, if we are interested in the relationship between A_t and A_{t+2}, we can write:

$$A_{t+2} = A_{t+1}(1 + r) = A_t(1 + r)(1 + r) = A_t(1 + r)^2$$

If, for instance, the rate of interest is 10 percent, then $100 today becomes $100(1 + 0.10)^2 = 121$ in two years, and $21 dollars is the compensation or premium for giving up control over $100 of present purchasing power for a period of 2 years.

We began this exposition with a brief reference to the agony of having to wait for our pleasure, and now we are talking about how money grows when used to buy a bond or deposited in a savings account. The connection between these two lines of thought is roughly as follows. The interest rate is an objective criterion. If you visit your local bank or brokerage office, you can be quoted the interest rate on such and such a type of deposit, or a security issued by a certain corporation or public authority. Let us assume that this interest rate, or *yield,* is 10 percent regardless of what type of account or security is involved. At the same time the individual making the inquiry will possess a subjective preference or subjective discount rate concerning his or her willingness to surrender money today in return for money in the future. Some people might consider 10 percent a perfectly satisfactory reward for waiting; thus they will be inclined to put a sizable portion of their salary in a savings account offering this rate of interest. The subjective discount rate of others might be so high that they would never consider postponing any consumption unless the rate of interest on their savings was at least 100 percent.

Unfortunately, however, neither economists or psychologists have had much luck measuring individual discount rates and no luck at all in aggregating them into a kind of community discount rate. Instead, the practice has been to use interest rates as a proxy for aggregate subjective discount rates, and so these two terms are often used interchangeably in the literature of economics. Thus a society with a low average rate of interest such as Switzerland might be regarded as a society where the average person has only a moderate preference for today's goods as compared to tomorrow's. For example, if the interest rate were zero, then Mr. or Ms. average saver could be thought of as someone who did not distinguish between consumption today and consumption tomorrow. However, this analysis, which is perfectly conventional in terms of mainstream economic theory, is not entirely correct. In most countries, and particularly Switzerland, individuals would have to save even if interest rates were negative in order to be able to exist during their retirement years and because of uncertainty concerning their future incomes.

Finally, if we receive various amounts of money at different times in the future, these future income flows can be discounted and summed to obtain a *present value*. As an example, we can inquire into the present value of $110 a year from now and $121 in two years. Assuming an interest rate of 10 percent, we obtain:

$$PV = \frac{110}{(1 + r)} + \frac{121}{(1 + r)^2} = \frac{110}{(1 + 0.1)} + \frac{121}{(1 + 0.1)^2} = 200$$

This concept can be generalized. Letting A_i represent a sum of money received in the ith period (with $i = 1, \ldots, n$), and with r the discount rate, or the interest rate if it is appropriate, we have as the present value of a stream of n payments:

$$PV = \frac{A_1}{(1 + r)} + \frac{A_2}{(1 + r)^2} + \frac{A_3}{(1 + r)^3} + \ldots + \frac{A_j}{(1 + r)^j} + \ldots + \frac{A_n}{(1 + r)^n}$$

The importance of this concept is that it allows us to rank income or payment streams on the basis of the size of their present value. At this point the reader should experiment with discounting and comparing various income streams with the same and different lengths, employing different rates of interest. One of the things that will be noticed is that as interest rates increase, present values decrease, assuming no change in the payment stream. This merely signifies a rise in impatience: A greater importance is attached to near as opposed to distant payments. Put another way, more remote satisfactions are downgraded. As a final observation the reader should recognize that flows in the form of the A's have been turned, via discounting and the interest rate, into a stock, which is PV. In other words, in terms of the previous relationship, there is no difference between having a sum (stock) of money PV at the present time, and n future payments (flows) $A_1, A_2, \ldots, A_n$.

Capital Values

Having looked at the elements of discounting, we can begin our examination of the theory of capital values. The issue here is a fairly simple one and, at least in this chapter, makes only a few analytical demands.

We have already established that in the presence of a positive interest rate, a given number of monetary units today is worth more than the same amount at any point in the future. For instance, with a 10 percent interest rate, $100 today can purchase a security or savings account that will be worth $110 a year from now. Conversely, $110 a year from now has a present value of $100. Still considering present values, let us consider the present value of a stream of receipts amounting to $100 a year, received at the end of each year of the next three years. As the reader can easily verify, the *PV* of this stream is 90.9 + 82.6 + 75.1 = $248.6, and in a world of perfect certainty this is what a rational individual would be willing to pay for this stream. The specification of perfect certainty should be carefully noted, since it ensures that anyone unsophisticated enough to pay more will be victimized by rational individuals who understand the concept of present value, while these same rational individuals would be unwilling to sell this particular stream of receipts for less.

Now let us turn from financial assets such as bonds or bank accounts to physical assets such as machines. The basic problem does not change. We are still interested in the present value of a stream of receipts, but in this case the receipts originate from producing and selling some kind of product with the aid of capital goods (that is, machines). However, the issue is complicated somewhat by the fact that labor and other factors of production usually cooperate with machinery; these factors must be paid; and moreover, physical capital has a tendency to deteriorate, or depreciate, over time. In addition, taxes must usually be paid on the sales or profits from the goods produced by these machines.

Before introducing depreciation and other problems, let us examine the unreal case where we have no labor or other costs, and we can buy a machine for $100 that produces goods that sell for $110 a year from now. Furthermore, assume that the machine depreciates completely after year 1. The cost of the machine is $100, while revenue or income from it is $110, and so the rate of return on the machine in this simple case is

$$(\text{revenue} - \text{cost})/\text{cost} = (110 - 100)/100 = 0.1 = 10\%$$

Note that nothing has been said in this example about where the money came from to purchase this machine—that is, we did not specify whether it was borrowed or whether it belonged to the purchaser. Later on we will show that in a textbook type of market, and perhaps in certain circumstances in the real world, the origin of these funds is unimportant. However, until this matter has been clarified, it will generally be specified whether the money used to obtain a physical asset is borrowed or obtained in some other way. (Note also that here and in the following paragraph I ignore operating costs.)

Next suppose that we have a rate of interest equal to 10 percent, and we borrow money to buy a machine that provides $120 in revenue after one year. Again assume that the machine depreciates completely after one year. In this example, repaying, or amortizing, our debt requires $100. Interest charges are $10, which leaves 120 − 100 − 10 = $10. Discounting this $10 back to the beginning of the first period gives the amount of profit, or increase in wealth, accruing to the person or persons acquiring the machine. Arithmetically this discounted value is equal to 10/(1 + 0.1) = $9.09. (Observe the terminology: in this case discounting is unambiguously equivalent to taking the present value.) The purpose of this calculation is to measure profit at the time when the machine was purchased. In other words, we want to place all revenues and expenditures on a common basis, and the procedure for doing this is to consider them at the same point in time. Thus $120 received a year from the reference date has, under the conditions of this example, a PV of $109.90 on the reference date. Since we have to pay only $100 for the asset that generates this PV, our profit is $9.09. An important observation will now be made: This would have been the profit regardless of whether we borrowed the money to buy the machine, or had the $100 to begin with. In the first case we end up with $9.09, while if we had the money we would end up with $109.09. The next thing to emphasize is that for an economist, a positive profit on a physical (or a financial) asset means that the asset gives a larger return than would have been available from an investment in a safe financial asset (such as a government bond). For example, with an interest rate of 10 percent, a $100 investment in a safe financial asset would have resulted in $110 after one year, instead of the $120 obtained in this example by investing in a machine. With a revenue of $110, economic profit is zero.

We can now consider such things as depreciation and taxes. In the preceding discussion, we assumed that the machine simply evaporated at the end of one period, leaving no scrap value. As things stand, the tax laws in most countries specify that physical assets that have a part in creating income, but which depreciate while doing so, can have a certain amount of their cost or purchasing price written off in each period as a business expense. In a world of taxes, this is an important provision. For example, consider the same one-period machine used in the previous example, but this time assume that operating costs for—say, wages and energy—come to $5, and assume a 25 percent tax rate on net revenue—where net revenue can be defined as gross revenue minus operating costs. The return to the owner of this machine is thus 120 − 5 − [0.25 · (120 − 5)] = $86.2. In other words, at the start of the period the purchaser of the machine had $100. He buys a machine and produces something from which he derives a gross income of $120. He pays operating costs of $5, and on a taxable income of $115, pays 25 percent ($28.75) in tax. He therefore ends up with $86.5 at the end of the period, which, discounted to the beginning of the period, gives a *PV* of $78.6. Profit is therefore negative (and equal to 78.6 − 100 = −21.4). Should he make the mistake of purchasing this machine, his wealth would diminish.

But had full depreciation been allowed on the capital asset, as would normally be the case, his taxable income would have been 120 − 5 − 100 = $15. Applying the same tax rate of 25 percent means that taxes would have amounted to $3.75. Aftertax income at the end of the period is thus 120 − 5 − 3.75 = $111.25. Using a discount rate of 10 percent, the *PV* of this amount would equal 111.25/(1 + 0.10) = $101.136. A purchase price of the machine of $100 would indicate a positive profit and increase in the wealth of the asset holder of $101.136 − 100 = $1.136. This amount is less than the increase that would have been realized had revenues been untaxed; but, as shown, it is preferable to the situation that would have existed had there not been a depreciation allowance.

We can continue this discussion with a few simple extensions. If we compare two assets—for example, two machines or a machine and a bond—discounting in both cases must refer to aftertax income. Suppose that the income from the safe bond or financial instrument referred to earlier is taxed at the rate of 20 percent. Then, with an interest rate of 10 percent, a $100 bond yields $8 a year instead of $10. As a result, the effective interest rate—that is, the interest rate used in discounting—should be 8 percent instead of 10 percent. In addition, as the interest rate goes down, the *PV* has a tendency to go up—at least in the simple examples that we are dealing with here. But as the literature of capital theory makes clear, this effect is illusory because of a phenomenon known as *reswitching*.

Some question can now be raised as to whether some market rate of interest should always be used in discounting. As noted earlier, the rate of interest differs from the discount rate as a concept in that the first is objective (due to daily quotations by banks and other financial institutions), while the discount rate is subjective in that it depends on each individual's weighing of present against future pleasures—or, in the jargon of formal economic theory, present against future utility. It is thus quite conceivable that income streams are discounted by subjective values without the slightest relation to market interest rates. Certain hard-driving businessmen have been known to use a 20 percent discount rate on income streams when the rate of interest on high-class bonds or bank accounts was 7 or 8 percent; and when asked why, they stated that they were not interested in projects that did not guarantee a return larger than

15 percent. On the other hand, analysts for such projects as dams and highways, and economists employed by public utilities, sometimes use discount rates that are lower than the market rate of interest since they are concerned with something other than making a profit. This important issue is examined in the following example.

Suppose we have a coal mine with two grades of coal, good and inferior. With a little care, only the good grade of coal need be mined, and as a result the mine can produce 1,000 bundles of coal in one year that can be sold for a profit of $1 per bundle. This practice, which in some contexts is referred to as "high grading," shortens the life of the mine. For the purpose of this example it will be assumed that if high grading takes place, the installation must be closed down after one year. But if a carefully chosen amount of the inferior coal were also extracted with the good grade of coal, the mine could last two years. Given the latter case, let us assume that 550 king-sized bundles per year comprising a blend of both good and inferior coal can be extracted, with each bundle again selling for a profit of $1 per year. Next let us make the assumption that employment per year is the same whether we produce 1,000 normal bundles or 550 king-sized bundles. Thus if the mine functions for two years, we will have twice the employment that we would have had with a one-year operation.

Now let us suppose that the owner of the mine is an overachiever with a deep aversion to accepting the average rate of profit. He thus decides to use a 30 percent discount rate when considering the two profit streams that he can obtain from his mine. The authorities, however, are dissatisfied with this decision. They say that it will shorten the life of the mine, causing employment problems and wasting the resource represented by the inferior grade of coal, which, even though inferior, is still potentially salable. They suggest the use of a 10 percent discount rate since, according to them, a project (or *prospect*) that registers a positive profit (or an increase in wealth) when a 10 percent discount rate is used is still making a decent profit. Furthermore, to keep things simple, let us assume that the mine owner received the mine free, and thus any positive PV means a positive profit. Accordingly, the only problem facing this is choosing between alternative profit streams, and choosing the largest.

Figure 2–2 shows a calculation of *PV*s for the two profit streams, using both the mine owner's preferred discount rate and that of the authorities. As shown, the mine owner receives his profits at the end of the year, and the values of these profits discounted to the beginning of the first year are shown in parentheses.

As the calculations in figure 2–2 show, a 30 percent discount rate rejects the two-period scheme because its *PV* is lower than a one-period

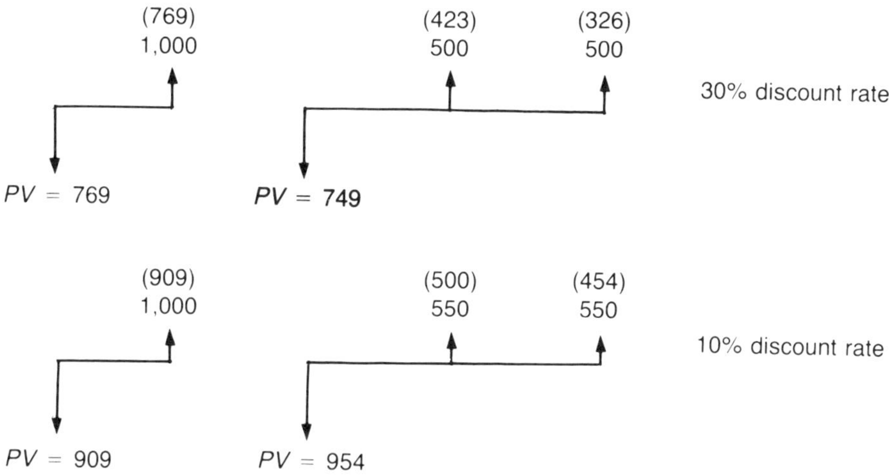

Figure 2–2. Two Profit Streams from a Coal Mining Operation Discounted at Different Discount Rates

operation. On the other hand, with a 10 percent discount rate the two-period operation is the most preferable. This type of arrangement has led to the use of the expressions *social* and *private* discount rates, where in terms of the preceding example the social discount rate would be that chosen by the government on the theory that the smaller the discount rate, the more weight given the future. On the other hand, a 30 percent discount rate had the effect of cutting off the future.

This example may appear to be trivial, but it covers several very important points that all readers (and particularly students of economics) would do well to examine. One of these is that where coal and other minerals are concerned, the discount rate is a key factor in determining whether the best grade of the mineral should be extracted before any of the inferior grades are removed. Many theoretical results exist, however, which suggest that the best grades should always be extracted first. In addition, shortening the life of the deposit through extracting the best grade first is a well-known occurrence to many mining engineers working in the coal industry. What happens is simply that the inferior grades are mixed with waste and become so diluted that, given the present state of the mining arts, it is not worthwhile to attempt to recover them. An outstanding discussion of these issues is found in Folie and McColl (1978).

Appendix 2A
Units and Conversion Factors, Time to Exhaustion, and Discounting

Units and Conversion Factors

As noted in the beginning of chapter 2, 1 t is the designation of one metric ton, or 1 tonne, which equals 2,205 pounds. We also have a short ton, which in most countries is simply called a ton. One short ton (1 ton) = 2,000 pounds; thus 1 t = 1.103 tons. Finally, there is a long ton, which is 2,240 pounds.

When working with energy, we are often interested in heat, which in economics textbooks and the press tends to be measured in British thermal units (Btu), where 1 Btu is the amount of heat needed to increase the temperature of 1 lb of water by 1 degree (1°) Fahrenheit. One metric ton (1 t) of bituminous coal has, on the average, an energy content of 26,460,000 Btu, whereas 1 t of crude oil has an energy content of 42,560,000 Btu. One thousand cubic feet (1 kcf) of natural gas has an energy content of about 1,000,000 Btu. These, of course, are average values.

Prefix	Symbol	Power	Meaning	Example
kilo	k	10^3	thousand	kW (kilowatt)
mega	M	10^6	million	MW (megawatt)
giga	G	10^9	billion	GW (gigawatt)
tera	T	10^{12}	trillion	TW (terawatt)
peta	P	10^{15}	thousand-trillion	PJ (petajoule)
exa	E	10^{18}	million-trillion	EJ (exajoule)

The joule (J) is widely used in the physical sciences as a measure of heat energy, and it is said that it is only a matter of time before it completely replaces the Btu. The kilowatt-hour (kWh) is a very popular and well-known unit for measuring energy in electrical form, and is discussed in some detail in chapter 7. A less common unit is the therm (100,000 Btu = 1.055×10^8 J), and in the United States the quad (10^{15} Btu) is often used. A short table of energy equivalencies follows:

	Joules (J)	kilowatt-hours (kWh)	Btu
1 joule	1.0	0.278×10^{-6}	0.948×10^{-3}
1 kWh	3.6×10^{6}	1.0	3.412×10^{3}
1 Btu	1.0555×10^{3}	0.293×10^{-3}	1.0

It is worth remembering that 1 million tonnes of oil (equivalent) can be converted to British thermal units should this unit be relevant to the discussion. For this transformation we need to know that 1 t = 7.33 barrels (bbl), and 1 bbl = 5,800,000 Btu on the average. The most popular unit for measuring both the consumption and production of oil is barrels per day (bbl/d). For example, OPEC oil production at the present time is approximately 18 million barrels of oil per day (18 Mbbl/d). This can be turned into another well-known unit, millions of tonnes per year (Mt/y), by multiplying by 50.

Power is defined as the rate of doing work, another concept that will be taken up in detail in chapter 7. Perhaps the best known unit for measuring power is the horsepower (hp), but in measuring electrical power the watt (W) is almost universally used. The following table of equivalencies is useful.

	watts (W)	Horsepower (hp)	Btu/hour
1 watt	1.0	1.341×10^{-3}	3.41
1 hp	0.746×10^{3}	1.0	2.54×10^{3}
1 Btu/hour	0.293	0.393×10^{-3}	1.0

Finally, in summary form, some useful equivalencies are:

1. 1 barrel crude oil = 42 U.S. gallons, and weighs 0.136 metric tons (0.136 tonnes)
2. 1,000 cubic feet (1,000 kcf) natural gas = 28.3 cubic meters, where 1 cubic meter = 35.33 cubic feet = 35.33 ft^3
3. 1 kilowatt-hour (1 kWh) of electricity = 3,411 Btu = 860 kilocalories
4. 1 metric ton of coal *equivalent* (1 tce) = 27,783,000 Btu, where a ton of coal equivalent is a metric ton with a heat value of 12,600 Btu/lb
5. 1,000 cubic feet of natural gas = 1.035×10^{6} Btu = 2.61×10^{8} calories
6. 1 tce = 4.8 barrels of crude oil, on the average.
7. 1,000 cubic feet of natural gas = 0.178 barrels of crude oil
8. 1,000 kilowatt-hours of electricity is equal to 0.588 barrels of crude oil.
9. 1 centimeter = 0.032808 feet = 0.3937 inches
10. 1 foot = 0.3048 meters; 1 inch = 2.54 centimeters

11. 1 gram = 0.00220462 pounds (avoirdupois); 1 kilogram = 2.2046223 pounds
12. 'e' = 2.7128 (the base of the natural logarithm system)
13. One U.S. dollar = 1,000 mills, or 1 cent = 10 mills

Time to Exhaustion

The next topic concerns the effect of the growth rate of mineral use on the time to its exhaustion. If we take g as the growth rate for the mineral, and X_t as the consumption (or production) of the mineral during year t, we have:

$$X_t = X_o e^{gt}$$

The term X_o is the consumption of the resource in some arbitrary initial year. Cumulative resource use is then defined as:

$$X = \int_o^T X_t dt = \int_o^T X_o e^{gt} dt \quad \text{or} \quad X = \frac{X_o}{g}(e^{gT} - 1)$$

With $\overline{X}$ of the resource available, we can rearrange this expression to get, as the years to exhaustion T_e:

$$T_e = \frac{1}{g} ln\left(\frac{g\overline{X}}{X_o} + 1\right)$$

It is also interesting to observe the effect of changes in $\overline{X}$ on T_e. Differentiating we obtain:

$$\frac{dT_e}{d\overline{X}} = \frac{1}{g\overline{X} + X_o}$$

What we see here is that substantial changes in $\overline{X}$ are not reflected in the time to exhaustion. For instance, in chapter 2 an example was constructed in which the amount of the resource increased by a factor of 10, but the time to exhaustion increased only by a factor of 2.3.

Discounting

We can now continue, on a somewhat more advanced level, the discussion at the end of chapter 2 that dealt with the present value of an income stream. Take a situation where we have A_o as present income, A_1 the income at the end of period 1, A_2 at the end of period 2, and so on. Assuming a constant rate of interest (or discount rate) r, we define present value as:

$$PV = A_o + \frac{A_1}{(1 + r)} + \frac{A_2}{(1 + r)^2} + \ldots + \frac{A_n}{(1 + r)^n}$$
$$= A_o + RA_1 + R^2A_2 + \ldots + R^nA_n$$

where we have defined $R = 1/(1 + r)$. If we also, for the sake of simplicity, assume $A_o = A_1 = \ldots = A_n = A$, the previous expression can be simplified to:

$$PV = A(1 + R + \ldots + R^n)$$

Multiplying both sides by $(1 - R)$ we get:

$$\begin{aligned} PV(1 - R) &= A(1 + R + R^2 + \ldots + R^n)(1 - R) \\ &= A(1 + R + \ldots + R^n) \\ &\quad - A(R + R^2 + R^3 + \ldots + R^{n+1}) \\ &= A(1 - R^{n+1}) \end{aligned}$$

Thus:

$$PV = A\frac{1 - R^{n+1}}{1 - R}$$

Usually we have $0 \leqslant R \leqslant 1$, and thus if $n \to \infty$, we must have $R^{n+1} \to 0$. In this situation the previous expression becomes:

$$PV = \frac{A}{1 - R}$$

This expression resembles the well-known multiplier, but by using $R = 1/(1 + r)$ we get $PV = (1 + r)\, A/r$. This is not a very familiar equation in the present context, nor should it be. The usual arrangement is to make an investment at time t_o, however, the revenue from this investment does not begin at t_o but one period later, at t_1. Thus in the above derivation we should have $A_o = A = 0$, and so for our PV in the case of an infinite income stream we have:

$$PV = \frac{A}{r}(1 + r) - A = \frac{A}{r}$$

This, of course, is much more familiar.

Before leaving this topic it should be noted that later in this book the concept of capital cost will arise. On the most elementary plane the capital cost can be thought of as the price paid for the use of physical capital over a definite period of time. Without depreciation of the asset, and with all else remaining the same, the capital cost would reduce to the interest rate; but in multiperiod situations of the type being discussed in this section, where there is depreciation of the asset, capital cost might best be considered as the uniform level of return the firm must earn to achieve a net PV of zero—that is, a zero economic profit. Using a continuous representation, and taking the price of the asset as unity, we get:

$$1 = \int_o^T xe^{-rt}dt = \frac{-x}{r}(e^{-rT} - 1)$$

This expression can be rearranged and put into discrete form to obtain an equation that will be derived in another way in chapter 7:

$$x = \frac{r}{1 - e^{-rT}} = \frac{r}{1 - \frac{1}{e^{rT}}} = \frac{re^{rT}}{e^{rT} - 1} \quad \text{(or) } x \approx \frac{r(1 + r)^T}{(1 + r)^T - 1}$$

3
Demand, Supply, and Price

The problem with coal is, at bottom, extremely simple. Nobody, except the coal industry, really wants a return to the age of coal at the present time. True, immediately after the first oil price shock, coal did hit the comeback trail. In the United States, production increased by 20 to 25 percent in a few years, as did the consumption of coal by utilities; and coal exports increased rapidly. In the same vein, expectations were that the second oil price shock would result in an acceleration in the use of coal; but the worldwide recession following this event led to a very large fall in the demand for coal since, as pointed out in chapter 1, there was a dramatic fall in the demand for all energy materials. Furthermore, for the United States, competition intensified rapidly on world coal markets, because South Africa and Poland are inherently able to supply coal to Europe at lower prices than the United States, and the same is true for Australia (and South Africa) where Japan is concerned. Coal consumption is still expected to grow almost everywhere during the remainder of this century, but anticipated growth rates are being constantly scaled down. In the case of the United States, coal consumption is very largely a function of what takes place in the electric utility sector, where 70 to 80 percent of the production of thermal coal is used; but with present forecasts of the growth in electricity consumption averaging only about 2.5 percent (as compared to the 7 to 8 percent of a decade ago), coal's largest market appears to have lost its bloom.

Even so, the U.S. coal industry is rocked with explosions of euphoria from time to time, and perhaps some of this optimism is justified. The demand for energy will undoubtedly rise again at a fairly impressive rate, and when it does proportionally more coal should be demanded than oil. A major obstacle to a rapid expansion in the use of coal is the slow replacement rate of the existing stock of oil- and gas-fired boilers; but should the world economy recover, more investment in boiler conversion might seem attractive and the rate of growth of electricity demand might pick up slightly. (Although boilers can be designed to burn any type of fuel, oil and gas are easier to burn than coal, because unlike solids these fuels mix readily with air during combustion. Coal has a lower heat content than oil or natural gas, more is required to raise the same amount of steam, coal handling problems are more complex, and coal consists largely of noncombustible ash—which complicates its use in boilers.) Finally, real progress is being made in developing coal combustion equipment that can burn coal in an environmentally acceptable way (see the conclusion of chapter 6). On a more abstract plane, as pointed out by A. Denny Ellerman of the U.S. National Coal Association, steam coal is the natural backstop for oil in the sense that there is so much of it that its price does not include a scarcity premium or royalty. (The backstop concept will be explained in appendix C.) As long as the world's coal-producing capacity continues to expand at the rate experienced over the last decade, coal will theoretically be able to set an upper limit on the oil price for a long time in the future.

But all this is for tomorrow, and not today. Today, even if coal producers in the United States and Europe are not going bankrupt, they are not making a great deal of money. What they as a group are doing instead is borrowing heavily to make premature or unjustified additions to coal-producing, -handling, and -transporting capacity. This kind of behavior means that the present coal glut might persist even when consumption rises. In the long run this type of irregularity has a way of curing itself, but until then the coal industry might have to accept an intolerable economic waste in unused or misused resources that neither it nor the industrial world can afford any longer.

The U.S. coal industry provides a fascinating example of hopes raised and, to a considerable extent, dashed; and perhaps also a lesson in resource misallocation. There are actually two U.S. coal industries: one in the East, which is capable of producing about 650 million tons of coal a year; and one in the West, with an annual capacity of about 325 million tons. For a long time it was generally considered that the underground mines of the Appalachian region—where coal seams average 5.3 feet, and which are worked by miners kneeling or lying on their sides—would be unable to compete with the open pit mines west of the Mississippi, where seams can be up to 90 feet thick, and machines scooping coal at the rate of 100 tons per worker-day can be employed. Moreover, western coal is low sulfur coal, almost in the same sulfur content class as the low sulfur coal of Appalachia. But eastern coal lies close to the major industrial centers of the U.S. Northeast and to seaports facing Europe. Since coal has a low value relative to weight, long journeys tend to make it uneconomical to mine unless transport costs are low. Consequently, the upturn in world demand that took place in the middle of the 1970s meant a boom period for Appalachian coal where, among other things, many small mines expanded production as fast as they could. It could be argued, however, that the expansion in output should have taken place in larger mines where costs were lower and economies of scale might have been realizable. Had the monopoly profits been eliminated from railway rates—particularly those of lines leading to the Gulf Coast—then U.S. coal would have been able to compete in Europe, and in theory everyone might have been better off (including the railroads, which, in the long run, would have won in volume what they lost in price. It is also possible that a scheme could have been designed to provide railroads with a subsidy from the profits of coal companies—had any profits been realized.) It was at approximately this time the claim was made that Appalachian coal could sell cheaper in Western Europe than the coal that Europeans could obtain from local mines, and on the average this may have been true; but it is more useful to know that during this period several thousand Appalachian firms were involved in coal production. For instance, in Kentucky, the most important coal state, 950 comparatively small mines produced 60 percent of that state's coal in 1980, with 750 of these mining less than 100,000 tons.

The great number of coal producers in the United States, and the rapidity with which large numbers of individual producers were expanding production, quite simply meant that when the demand for coal fell after the second oil price shock, it was impossible for many producers to maintain satisfactory margins. Moreover, unlike the textbook algorithm of perfect competition, output did not adjust until substantial losses were common throughout the industry. By the time overall production was reduced, prices were so low that many producers who could have survived had high-cost producers decreased production earlier, or railroad rates been more flexible, were forced into bankruptcy. Something that should be appreciated here is that had most of the expansion in output been accounted for by low-cost (but high-volume) producers in the United States, the situation might have remained favorable for the U.S. coal industry on the whole. Not only would these large producers have been better equipped to compete with low-cost coal from Poland and South Africa, but the long-term contracts that could have been signed by producers in possession of huge supplies of coal might have been of such a duration as to preclude price falls due to an increase in market supply and minimize the possibility of the supplier having to accept lower prices because of a fall in demand.

European coal producers, who are much smaller in number and enjoy an easier access to subsidies than their U.S. counterparts, have managed to maintain their production until recently. The general economic malaise in Western Europe is so deep, however, that major production cuts have become unavoidable. Both the National Coal Board (England) and Charbonnages de France seem to have concluded that coal output in their countries must be reduced to a level that makes economic sense. In Germany, Ruhrkohle and Saarbergwerke are now attempting to reduce their production by about 15 percent. Despite his promise to the coal miners of France when he was their candidate for president, President Mitterand intends to reduce the French mining force by almost 40 percent as the switch to nuclear energy is made definitive in that country. What these cuts will most likely mean is that South Africa, and perhaps Poland, will increase their shares of the Western European coal market.

The Demand for Coal

The demand for coal is a derived demand in the sense that coal is an input in activities for producing final goods and services (such as heating and lighting). As already indicated, the principal use of steam (or *steaming* or thermal) coal is for electricity generation, as table 3–1 also makes clear; but a huge demand still exists for coking (or metallurgical) coal, which is used to make steel. Before 1974, the world trade in coal was almost entirely a trade in coking coal, and the great fear at that time was that the world was gradually running out of this commodity. However, as a share of the total coal traded or used, it is expected that the relative demand for coking coal will shrink a great deal in the future.

Some explanations of table 3–1 are perhaps in order. The conversion of coal into gas (and selling it in that form) is perhaps an ancient process. In the early years of this century, many cities and towns (including

Table 3–1
Use of Coal, 1975 and 1979
(thousands of tons)

	1975	*1979*
Miscellaneous		
Power plants	511,282	665,728
Coking	273,208	246,639
Coal gas production	8,592	6,128
Briquettes	6,430	6,442
Direct consumption	4,720	3,366
Transport sector	1,465	993
Industrial sector		
Iron and steel	7,768	8,451
Chemical	5,452	5,312
Other	76,857	75,700
Other sectors	34,051	31,502
Total	928,181	1,049,678

Source: International Energy Agency, 1974, 1975, 1981; Forskningsgruppen fõr Energisystemsstudier, Stockholm University, 1983.

virtually every major city in the eastern United States) had gasworks where gas was manufactured from coal for the purpose of lighting and cooking. In some contexts it was customary to refer to this gas as "town gas." The entry next to iron and steel is coal used to provide such things as power and light for steelmaking installations. Almost all industrial installations require a great deal of energy, and one of the reasons the countries of the Middle East expect to do so well with their refineries is that inexpensive gas cannot only be used as an input for the production of chemicals, but also to provide the energy needed to run these refineries.

In discussing the demand for coal, we are led immediately to speculate on the elasticity of demand for coal, where this elasticity is defined at the percentage change in the demand for coal, given a 1 percent change in the price of coal. As far as I know, there are no authoritative estimates of the elasticity of demand for coal, and unless I am mistaken the reason for this is similar to the reason for there being no useful elasticity estimates for petroleum until recently. Before the 1973–1974 oil shock there was hardly any variation in the coal price, and therefore it was impossible to use regression techniques to estimate these elasticities. At the same time, however, we do know that finite own elasticities of demand exist for coal, even if we do not know their value; and we also know that substitution elasticities exist for coal vis-à-vis oil and gas and perhaps also nuclear energy. In considering the matter of substitution elasticities, it seems clear that if the relative price of oil to coal should exceed a certain level, and give some indication of staying there, there would be a gradual phasing out of oil-fired electricity generating capacity in favor of coal as the oil-fired capacity depreciated; and at the same time there would be a speed-up in the conversion of boilers from gas and oil to coal. Although I am not prepared to defend any estimates of the own or substitution elasticities of demand for coal at the present time, some back-of-the-envelope calculations that I made recently indicate that these elasticities are somewhat higher for coal than for oil—at least for the United States. It is possible, though, that these comparatively high elasticities can be explained by extraneous factors, such as the disfavor that has fallen on the U.S. nuclear industry.

By way of contrast, something that would tend to reduce the substitution elasticity of coal for oil is the quite readily observable tendency of the price of coal to follow the price of oil, as can be seen in figure 3–1. Thus a switch from oil to coal is not always a guarantee that a consumer is going from an expensive to an inexpensive energy medium. What must be understood here is that the coal industry in many countries is not inherently a low-cost industry. Coal-mining unions are, by tradition, among the most militant in the world, and in those situations in

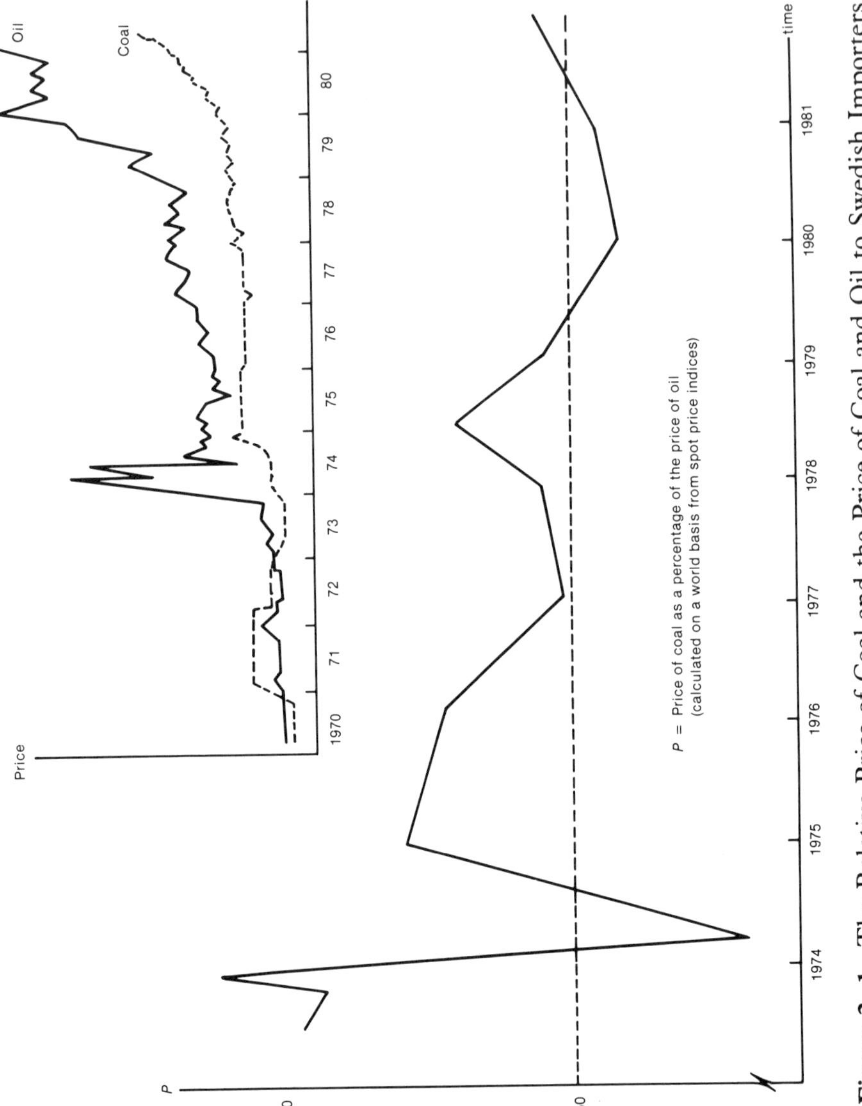

Figure 3–1. The Relative Price of Coal and the Price of Coal and Oil to Swedish Importers

which coal miners are extremely productive, as in the nonunionized sector of the U.S. coal industry, they are very well paid. In 1983 wages among nonunion employees in the U.S. coal industry averaged about $27,500 per worker. The question that nobody can answer however is how rapidly wages and salaries can adjust should productivity turn down again and the demand for coal remains low. Productivity fell 32 percent in the U.S. coal industry between 1969 and 1979 before making an impressive recovery in 1980. Some observers have suggested that this recovery was a false dawn, but others insist that a younger work force, together with such innovations as production bonuses, will eventually bring about a fundamental change in the coal industry.

Western European consumption patterns indicate that only West Germany and Turkey produce enough coal to satisfy domestic demand. Britain is an importer despite its considerable overcapacity, because South African and U.S. coal can be landed and sold in Britain cheaper than a large percentage of the domestic product. The United Kingdom and West Germany together produce more than five times as much coal as the remainder of Western Europe, where the only other major producers are Belgium, France, Spain, and Turkey. These last two countries have plans to raise production; however, it should be pointed out that their coal industries are not modern and require a great deal of investment if they are to increase production by the amounts being discussed. Spain is particularly interested in raising the amount of coking coal being produced domestically, while in Turkey lignite production for local use is scheduled to increase.

France has the third largest coal consumption and production in Western Europe, and at one time President Mitterand declared that coal should be mined as long as it was cheaper than importing an energy equivalent of heavy fuel oil. France is also the largest importer of coal in Western Europe at the present time, but there is considerable doubt as to whether the intensity of coal usage in that country can be maintained given its plans for expanding nuclear energy and the increased availability of natural gas from Algeria and the Soviet Union. According to Sten Kjellman (1981), developments in Belgium will probably follow the same pattern as in France.

Italy and Denmark import a considerable amount of coal and also have plans to expand the use of coal. Denmark is reluctant to use nuclear energy, and tentative plans have been drawn up to base the country's energy economy on coal. As it happens, though, environmental considerations are extremely important in Denmark, and if large amounts of natural gas became available at a reasonable price, it is possible that an attempt would be made to freeze coal consumption at about its present level. Similarly, very large natural gas pipelines from Algeria and the

Soviet Union now terminate in Italy, and with the price of natural gas falling, the present Italian government has indicated that less coal will be consumed in the future than originally planned.

Norway, of course, with its huge supplies of oil and gas, sees steam coal only as a competitor, and has no intensions of using steam coal in the domestic economy in the near or distant future. Finland consumes a certain amount of steam coal, but at the present time has access to so much Russian oil (and potentially to gas) that it seems unlikely that it would be interested in buying more coal. In addition, there are no political barriers to an increased use of nuclear energy in Finland, and the Finnish nuclear sector is very efficient. The Netherlands plans a modest increase in the use of coal, and here it is interesting to point out that Holland may possess some of the highest quality coal in Europe, although apparently this coal is offshore under—or in the vicinity of—the Groningen gas fields.

With the price of both oil and gas falling in Europe, it is difficult to make a reliable prognosis of the future of Western European coal demand. Unless I am mistaken, however, coal is going to meet much harder competition from natural gas than was considered likely just a few years ago. Soviet gas is now definitely in the Western European energy picture, and one thing it will probably continue to do, just as it has been doing for the last year, is to press down the price of all natural gas and to a certain extent the price of coal. In fact, with natural gas selling at a crude oil equivalent price of $25 per barrel—and capital costs for natural gas installations well under those of coal-fired units—many observers have started to suggest that more gas should be used in the industrial sector, where earlier natural gas was regarded as less economical than coal. Although some people do not think so, I believe that simple economic logic calls for giving the Soviet Union the opportunity to deplete its huge reserves of natural gas at a much higher rate than presently conceived in Western Europe.

Finally, some mention should be made of interfuel substitution. In the not too distant past, crude oil was the premium fuel in the industrial sector, but it seems unlikely to retain this distinction. Table 3–2 presents some historical data, as well as some World Bank projections. Nuclear-based electricity and natural gas are envisaged as the main substitutes for petroleum in the residential sector, while coal and nuclear-based electricity fulfill this function in the industrial sector.

A point of particular importance in table 3–2 is the part that oil plays in the energy requirements of the LDCs. Even in the year 2000 oil plays a dominant role in the energy demand of these countries, and given their rate of population increase, it appears likely that this will be the main source of the upward pressure on the oil price at that time.

Table 3–2
Fuel Shares in the Industrial and Residential Sectors
(percents)

	Industrial Countries					*Less Developed Countrioes*				
	1970	*1978*	*1985*	*1990*	*2000*	*1970*	*1978*	*1985*	*1990*	*2000*
Industrial sector[1]										
Crude oil and gas liquids	41.3	44.8	39.6	38.0	34.0	51.8	57.5	48.4	47.1	43.9
Natural gas	24.7	21.8	19.3	18.5	17.6	16.8	15.8	19.8	18.5	18.2
Coal	20.2	17.6	22.3	24.1	27.8	21.6	15.8	20.9	23.5	28.2
Electricity	13.8	15.8	18.8	19.4	20.6	9.8	10.9	10.9	10.9	9.7
Residential sector										
Crude oil and gas liquids	46.5	41.6	30.9	28.4	23.4	62.5	57.1	44.1	41.0	36.0
Natural gas	28.5	32.2	36.1	35.8	35.8	4.7	6.4	13.9	15.0	18.1
Coal	9.5	4.2	3.0	2.1	1.1	15.1	17.1	15.2	12.9	9.3
Electricity	15.4	22.0	30.0	33.8	39.7	18.3	19.4	26.7	31.1	36.6

Source: World Bank publications; International Energy Agency: Energy Statistics, 1982.

[1]Including metallurgical coal.

Supply and Price

On the supply side of the market, it is quite clear that in the coal industry supply curves rise to the right in the conventional manner. Price increases cause increases in supply either through existing mines expanding production or restarting operations, and in the long run new mines will come on stream. This industry is extremely heterogeneous, with mining firms that own rich deposits making high profits at a time when less favored firms are just earning enough to stay in business; nowhere is this more true than in the United States. In 1980, the U.S. coal industry's most profitable company was a medium-sized firm, MAPCO (of Tulsa, Oklahoma), whose aftertax earnings came to $5.77 per ton; and it appears that in the coal industry the position of the majors is not at all comparable, profitwise or otherwise, to the position of the majors in the oil industry. Figure 3–2 provides some supply schedules constructed by the U.S. Department of Energy for Australia and the United States.

In figure 3–2, the supply curve for the Gulf Coast should fall over

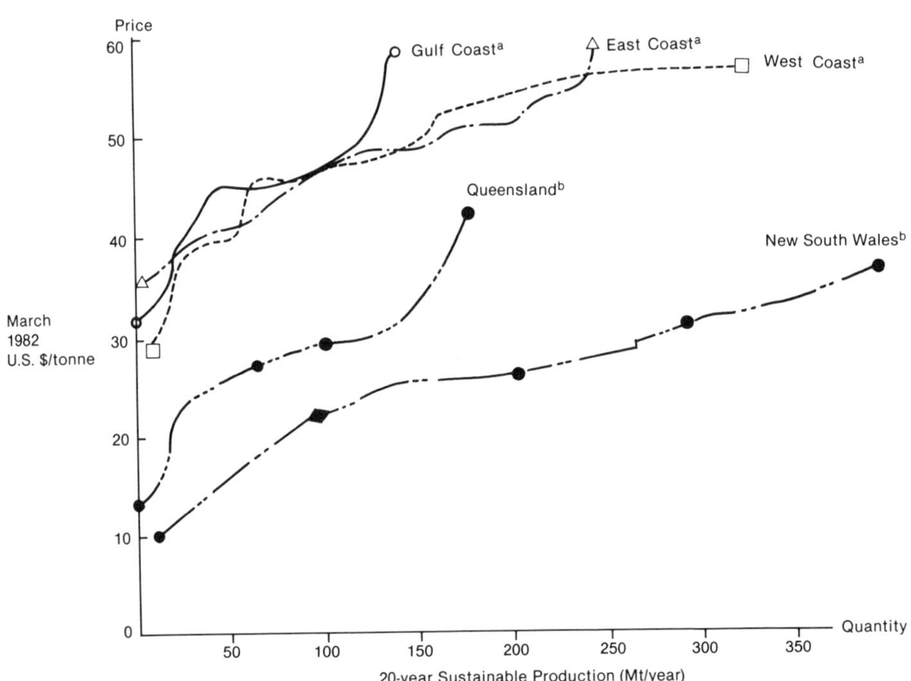

[a]Intermediate-Sulfur Steam Coal
[b]Low-Sulfur Steam Coal

Figure 3–2. Supply Curves for U.S. and Australian Coal

time because a larger percentage of U.S. production will be obtained from open pit operations in the western part of the country, where costs are considerably lower than in the East. Another factor influencing the displacement of this curve, as well as the other two for the United States, will be the development of inland transport costs; however, in my opinion it is only a matter of time before some arrangement is made with the railroads to reduce freight rates to an appreciably lower level. Also, not all coal mined west of the Mississippi is mined in surface (that is, open pit) operations. Proportionally, at least as much coal in the West can only be reached by underground mining as in the East. Also, today, most open pit mining is in the East, where output is twice that of the West.

As we know, supply curves are constructed on the basis of cost considerations, and the position of the Australian curves relative to those of the United States merely verify that the cost of exploiting deposits in these two countries are very different. Identifying and discussing these costs can be a tedious and complicated job, however, because a number of factors should be included in the analysis that cannot be meaningfully quantified in an elementary discussion of the economics of coal. Consider, for example, the very important matter of sulfur content. Coal with a high sulfur content can be turned into coal with a low sulfur content if it is properly washed. But all coal of a given sulfur content cannot be washed with the same ease, and thus the cost of obtaining coal with the desired sulfur content can vary greatly between mines and countries. Although it is not often realized, sulfur content is extremely important, because in general only low sulfur coal is traded internationally; also, as environmental regulation becomes increasingly stricter, buyers are constantly demanding ever-cleaner coals.

As to be expected, the most important elements in determining cost are wages, salaries, and productivity—at least for existing mines. Table 3–3 provides some indication of developments from 1963 to 1977. It would be nice to be able to extend this table past 1977, but the truth of the matter is that many coal statistics are not easy to obtain, and certainly the aggregation of statistics for individual operations into significant averages requires great sophistication. For example, a World Bank study published in 1984 relies almost exclusively on statistical materials published in or before 1977, the same year used as the statistical cutoff date of the prestigious World Coal Study (1980), whose wide ranging prognostications about the future of the international coal markets have unfortunately turned out to be considerably wide of actual developments. (What this means, incidentally, is that even persons with an almost complete knowledge of the coal industry can misinterpret trends, and that all forecasting exercises—regardless of how they are packaged—contain a large element of guesswork.)

As is apparent from table 3–3, productivity (as measured in output

Table 3–3
Average Productivity and Wage Costs per Ton of Coal in Four Important Coal-Producing Countries

	1963	*1969*	*1972*	*1973*	*1974*	*1975*	*1976*	*1977*
United States								
Production, hard coal	100	123	130	132	142	149	152	
Production value[1]	9.6	9.1	12.2	12.8	21.3	23.9	22.8	21.9
Production (ton/man-day)	14.5	18.1	16.1	15.9	15.9	13.4	13.1	13.5
Wage cost[1]	—	—	—	—	—	5.4	5.6	6.4
Australia								
Production, hard coal	100	176	226	232	241	253	276	294
Production, ton/man-year	1,516	2,597	2,934	3,001	3,022	2,834	2,986	3,074
Wage cost per ton[1]	—	—	—	—	3.9	4.3	4.4	4.4
South Africa								
Production, hard coal	100	124	139	149	157	165	182	204
Production cost[2]	—	—	—	4.9	5.4	5.9	8.3	—
Production (ton/man-year)	583	696	781	856	895	904	925	887
Wage cost[1]	1.75	1.97	1.74	2.2	2.4	2.4	2.7	3.1
Canada								
Production, coal	100	103	200	218	227	269	269	303
Production value[1]	13.9	14.1	14.5	14.7	18.4	27.0	25.7	25.0
Production (ton/man-year)	1,185	1,614	2,868	3,176	3,308	3,772	3,276	3,360
Wage cost[1]	6.2	5.4	4.5	4.1	4.1	3.9	4.5	4.3

Source: *Annual Bulletin of Coal Statistics for Europe* (various issues); *International Coal Report*, 1973–79; *United Nations Commission for Europe Bulletin*, 1975.

Note: Productivity for the United States given on the basis of raw coal production, while production for Australia and Canada is calculated on the basis of salable coal. Production of coal in index form (1963–100).

[1]In 1977 U.S. dollars.

[2]Estimated, excluding labor costs. In dollars/ton.

per time period) has increased especially fast in Australia and Canada. This is of particular importance for Australia, whose supply of exportable coal is much larger than that of Canada, at least at the present time. These figures show the deceleration in productivity in the United States mentioned earlier. In 1978 average production costs at eastern coal mines were between $20 and $30 per ton, while corresponding figures for western mines were $8 to $15 per ton. Furthermore, although new investment had all but ceased in the western part of the United States, marginal investment costs for new open pit mines were approximately one-half those of underground mines. In addition, on the average, only about 50 percent of an underground coal deposit in the United States can be recovered, while 80 percent—or more—of the coal in an open pit deposit can be extracted. In an underground mine, a great deal of coal is inevitably used in the structure of the mine (for the roofs, columns, and walls), though often somewhere outside the main seam or seams. It can easily happen that this particular coal becomes thoroughly mixed with waste, and so it does not pay to employ expensive factors of production to recover it while there are rich seams to tap elsewhere in the same deposit, nor does it usually pay to return to it later. It is also true that in the majority of cases, once a mine has been abandoned it will not be reopened to mine a comparatively lean deposit. (However, in those few instances in the United States where "longwall" mining systems are used, almost 100 percent of underground deposits can be removed. These systems protect workers with heavy steel canopies and are intended to cope with one of the two most serious production bottlenecks in underground mining: the absence of safe roof supports. The other bottleneck is the relatively slow speed with which mined coal is moved to the surface.)

Similarly, for open pit mines, it may be uneconomical to become involved with marginal ores that are distant from the center of the deposit, particularly if there are fairly rich deposits available elsewhere that can employ the services of the variable and fixed factors used in this kind of mining. New systems for underground mining will presumably involve automatic roof-bolting and temporary roof-support equipment; one coal company in the United States has been working on an arrangement to move coal to the surface in a constantly running water slurry. The large-scale employment of this kind of technology, which conceivably could result in a substantial boost in coal mining productivity, is still years away, and the simple truth is that no major advances in underground mining technology have taken place since the introduction of the continuous mining machine more than thirty years ago. On the other hand, it must be recognized that the coal mining industry was not regarded as having a particularly bright future until the first oil price shock. Price incentives have now appeared, or should eventually appear, that will result in important improvements in mining machinery and methods. It would also be helpful if labor relations improved, because in due course coal is going to be more valuable than ever.

South Africa and Poland

For pedagogical reasons, and because South Africa and Poland are the most important competitors of the United States in Western Europe coal markets, I now consider the supply situation in these countries. From a production of 54.6 Mt of coal in 1970, South African production advanced to more than 130 Mt in 1982, while exports increased by a factor of almost 15 to 30 Mt, and South Africa became the fourth largest exporter of coal. The real impetus for raising South African production was, of course, the first oil price shock, when the international market for steam coal suddenly came alive again. (At the present time, South Africa exports only 2 or 3 Mt of coking coal per year). Another important factor in the development of the South African coal industry has been the oil blockade against that country. Coal is now critical to maintain the high standard of living enjoyed by many people in that country. In fact, some South Africans want to impose fairly low limits on coal exports to conserve it for domestic use. Finally, racial policies in South Africa have meant the availability of a huge supply of inexpensive labor to dig coal, although it has been claimed that the wages of black South African miners are now rising faster than the wages of white miners—even if, on the average, they are still much lower.

There are somewhat more than seventy coal mines in South Africa, and for the most part mining is in the hands of private South African corporations. However, at least three large international oil companies—Royal Dutch Shell, BP, and Total (France)—are involved in foreign marketing. For instance, several years ago Shell was a partner in a 50-50 joint venture with Rand Mines, shipping several million tons of coal a year from Rand's Rietspruit mine to various foreign destinations, among them France. In 1979, BP, Total, and South Africa's General Mining Union Corporation, which is the country's second largest mining group (after Amcoal Ltd., a subsidiary of Anglo-American, which is also the world's largest producer of gold), opened the Ermelo mine, which cost more than $120 million to develop. Sale on the domestic market is handled through three producers' cooperatives—Transvaal Coal Owners Association (TCOA), Natal Associated Collieries, and the Anthracite Producers Association—as well as by the state-owned power and steel corporations ESCOM and ISCOR. Sasol, the state-owned synfuels company, has its own mines. Although the South African government loses no opportunity to describe itself as a champion of the free market system, coal prices are closely controlled and the authorities have stipulated that the selling price of South African coal must be commensurate with the ruling world market price. In addition, quantitative restrictions are placed on the export of coal by the issuance of export permits. The unwritten policy is that coal is too important to be left to businessmen. Much of the coal that is sold abroad is sold on contracts of five years or less. (Five-year contracts are sometimes called "evergreen" contracts.) Attempts have occasionally been made by foreign buyers to renegotiate long-term contracts—on one attempt, Japanese importers tried to force a $12 per tonne price cut on TCOA—but in general South African coal exporters are unsympathetic. Over the long run they know that their coal is the least expensive in the world, and they are prepared to use this knowledge in negotiating with—or not negotiating with—their customers and potential customers.

At the end of 1983 it appears that the average price of South African coal (f.o.b. Richards Bay) was $45 to $50 a tonne, although the present price is probably lower. Somewhat earlier, on the domestic market, coal was selling for $18 to $25 per tonne. It is difficult to estimate present production costs, but I put them between $20 and $25 a tonne. Accordingly, while domestic sales yield only a small margin if any, fairly high profits have been realized from foreign sales. The price at which South African coal sells abroad has been rising steadily because of the adjusting up of the price on long-term contracts. Still, South Africa is not averse to making occasional sales on the international spot markets, particularly in Europe, and sometimes South African spot prices have been in the bargain basement category. Of late the reason for these spot sales appears to be the inability of South African exporters to sell as much coal as they have been given permission to sell under the export license system.

Although South Africa exports to the entire world, Poland sells mainly to Europe. In 1979 only 3 Mt of total Polish exports of 41 Mt went to non-European buyers. Western Europe purchased 24 Mt, of which approximately two-thirds was steam coal and one-third coking coal. Internal political disturbances have occasionally caused a sharp fall in Polish output and exports, but the government regards coal exports as too important to the economy to allow them to stay depressed. Every effort is being made by the present government to maintain what it regards as the Polish share of foreign markets.

Most of the hard coal sold in Poland comes from three localities—upper and lower Silesia and the Lublin area. In addition to hard coal, a great deal of lignite is produced, and the large Polish reserves of brown coal are in the early stage of their depletion cycle. Present calculations are that the increased domestic use of brown coal will free increasing amounts of hard coal for the export market. Most of the production of hard coal takes place in underground operations, and as things now stand, productivity is not particularly high. What is needed in the Polish mining sector is a great deal of investment, but since the Polish government is operating on the rim of bankruptcy, and Polish foreign debts are so large that more private borrowing would be almost impossible to arrange, this investment is likely to be delayed by a good many years. Still, production is rising, and in 1983 coal accounted for $670 million of the $1.5 billion of hard currency Poland obtained from international trade.

At present, an internal struggle is taking place in Poland to bring the coal mines under more centralized direction. The Mining and Power Ministry is exploiting a loophole in the reform laws (passed in 1982) to keep decisionmaking powers from individual mines, and in the course of their manipulations this ministry has had the sixty-seven coal mines in Poland declared nonprofit-making utilities. The new competitive procedures for appointing managers have been overruled, and an attempt is being made to group certain types of companies into huge combines. For instance, the lignite industry is to become one huge combine that will include the producers of machinery for open cast mines.

The effect of all this on production and productivity cannot be judged at the present time, although it is useful to remember that the over-centralization of the Polish government under Edward Gierek was a main contributor to the inefficiencies that led to the near bankruptcy of Poland in the 1970s. The theory has been advanced that the Polish mining bureaucrats are deeply concerned about the power they would lose in the event of wide-ranging decentralization, and are attempting to regain the prerogatives they feel were taken away in the 1982 reforms. If this is true, they are fighting a losing battle, because in general centralization and economic efficiency do not seem to go together in Eastern Europe. As an indication of this, it can be noted that the weaker firms tend to greet the return to centralization with joy, since it will provide them with increased opportunities to conceal their inadequacies, and probably to obtain more subsidies.

Polish coal is sold abroad by a state organization called Weglokoks. This firm prefers to deal with just one importing body in each customer country, although for reasons that cannot be taken up here, this is impossible in West Germany. Most Polish coal is sold on short-term contracts (under one year), and Poland has been a diligent operator on the European spot markets.

The Inland Transport of Coal

Within the United States, between 65 and 70 percent of coal production is transported by rail. Almost 20 percent goes by barge or truck, while another 10 percent is used so near to the mine that it can be transported by conveyer belt or a similar arrangement. Much less than 1 percent moves through a slurry pipeline.

In terms of real resources, the cost of moving coal by rail is not excessive, particularly in built-up areas like the United States and Western Europe. True, a new spur line can cost a quarter of a million dollars per mile; but while a coal mine can take up to five years to open, a complete new railway line can be constructed in about a year. More important, with the railway system essentially in place, the main expense associated with raising the production of coal is not new lines, but more hopper cars for carrying coal and an upgrading of existing lines. At the present time these cars probably cost no more than $50,000 each; thus even if the very large number of 30,000 cars were added, a total annual cost of $150 million is probably insignificant when compared to the cost of constructing the utilities they would service. The number 30,000 was chosen here because this would be about the net annual investment in hopper cars required if transportation facilities in the United States were equipped to meet the coal production goals adumbrated by the International Energy Agency several years ago. However, given the present outlook for coal, the number is probably between 10,000 and 15,000.

The most efficient method for moving coal by rail is in a unit train, which consists of a large number of cars (often 100 or more) that carry coal directly between the producer and the consumer. These trains always operate as a complete unit and do not uncouple or stop except for loading, unloading, maintenance, or crew changes. This allows a very high annual car utilization level. In order to gain a deeper insight into economics of unit trains, let us examine a simple (and partially fictitious) numerical example dealing with the servicing of a large electricity generating facility. To begin, we can take D as the distance to be traveled between mine and utility, V as the average train velocity, L_d and L_u as the loading and unloading times. The time for a round trip is:

$$t = \frac{2D}{V} + (L_d + L_u)$$

For example, if D is 1,000 miles, V is 50 miles/hour, and L_d and L_u are 5 miles/hour, we get $t = 50$ hours.

Now we can ask about the coal requirements for the plant. Suppose that the plant generates 1,000 MW of electricity, where 1 MW = 1,000,000 watts = 1,000 kilowatts = 10^3 kilowatts. Thus, 1,000 MW = 10^6 kilowatts. As will be explained in detail in chapter 7, this is the power rating of the plant. The energy it would generate in 24 hours of operation is 24×10^6 kilowatt hours. As we see from the table of equivalents in appendix B, 1 kilowatt of electricity is equal to 3,412 Btu in a perfect system, but because of heat loss and other thermodynamic phenomena, perfection is impossible, and for the purpose of this example let us take 1 kWh = 10,000 Btu. Thus 24×10^6 kWh is equal to $24 \times 10^6 \times 10,000 = 24 \times 10^{10}$ Btu. (Again, see chapter 7 for a complete discussion of these concepts.)

The next step is equally simple, but very important. These Btu shall be obtained from coal, and to keep things simple let us assume that we have coal with a heating value of 10,000 Btu/lb. Using short tons (2,000 pounds), we have $10,000 \times 2000 = 2 \times 10^7$ Btu/ton. Thus 24×10^6 Btu can be obtained from 24×10^{10} Btu divided by 2×10^7 Btu/ton, which is equal to 12×10^3 = 12,000 tons of coal. Putting in the time unit, which was 24 hours, the requirements of this plant are 12,000 tons per day.

At this point we can go back to our unit train hauling coal from the mine to the utility, with each trip taking 50 hours. In 50 hours the utility will require $(50/24) \times 12,000$ = 25,000 tons. A typical hopper car carries 100 tons, and so in a 50-hour period 25,000/100 = 250 cars will be required, or 120 cars per day. If it is possible for a unit train traveling 50 mph to pull 120 cars, and if this utility operates at full capacity every day of the year, it requires a unit train a day to keep it supplied. Needless to say, a 1,000 MW plant would not operate at full capacity every day of the year, which means that fewer cars would be required; but by the same token a reserve of cars would be necessary because normally hopper cars are undergoing maintenance about 25 percent of the time. The final important factor is stockpiling capacity at the utility, because clearly when cars are available it might pay at times to deliver an excess of coal to the plant, stock it there, and then use it while some or all of the unit train is undergoing maintenance. All these things would have to be taken into consideration at the same time to determine the optimal number of hopper cars, but this problem falls within the purview of operations research and cannot be taken up here.

Appendix 3A
Exhaustible Resources and the Backstop Technology

Exhaustible Resources

Stripped of unnecessary detail, the exhaustible resource problem comes down to the following. Suppose we have a unit of a resource in the ground that can be removed at a certain cost MC_t, which can be interpreted as a marginal cost, and this unit can be sold for a known price p_t. It must be decided whether this unit should be removed. Because we are dealing with one unit at a time, we are interested in marginal values.

By removing the unit now and selling it at a price p_t, we obtain a *marginal profit* of $p_t - MC_t = MP_t$. If we put this marginal profit in a financial institution, our gain over one period is $MP_t(1 + r) - MP_t = rMP_t$, where r is the prevailing rate of interest.

If, instead of extracting it now, we leave the unit in the ground and extract it at the beginning of the next period, we realize a *capital gain* of $(p_{t+1} - MC_{t+1}) - (p_t - MC_t) = MP_{t+1} - MP_t$. The expression $p - MC$, which has been referred to as the marginal profit here, is sometimes called the net price. Now, if $rMP_t > (MP_{t+1} - MP_t)$, the profit-maximizing action is to remove the unit because the gain from selling the unit, banking the revenue, and withdrawing this money together with the interest on the revenue one period later exceeds the capital gain that would be realized by leaving the unit in the ground. Using the same logic, if $rMP_t < (MP_{t+1} - MP_t)$, then the unit is not extracted. Our equilibrium situation, which is the point at which we cease extracting the resource, comes when:

$$MP_{t+1} - MP_t = rMP_t$$

or:

$$\frac{MP_{t+1} - MP_t}{MP_t} = r = \frac{\Delta MP}{MP}$$

This last expression is a difference equation that, allowing time periods to become very small, can be written as a differential equation with the solution:

$$MP = MP(0)e^{et} \quad \text{or} \quad p - MC = MP(0)e^{rt}$$

In these expressions $MP(0)$ is formally the margin profit at time zero, although specialists in this type of analysis probably realize that the calculation of $MP(0)$ could, in the real world, involve some very sophisticated operations. That is, on the basis of expected demand and cost conditions in the future, an $MP(0)$ would have to be determined that ensured that the total quantity of the resource was used up at the end of the time horizon. In order to say a little more about this problem, let us take the special case where $MC = 0$. Thus in the previous result we get $p(t) = p(0)e^{rt}$ as the solution to $\dot{p}/p = r$. If $D(p)$ is the demand for the resource at price p, and $\overline{K}$ is the initial stock of the resource, then we must have:

$$\int_0^\infty D[p(t)dt = \int_0^\infty D[p(0)e^{rt}]dt = \overline{K}$$

In general this expression will provide a unique initial price $p(0)$, from which a price path characterized by a competitive equilibrium prevails.

Next, let us observe that the previous results can be obtained from the following Lagrangian expression, where the primary problem consists of maximizing the discounted profit $V(q)$ over a time horizon T (which could, of course, be infinite):

$$\overline{V}(q,t,\lambda) = \int_0^T V(q)e^{-rt}dt + \lambda(\overline{K} - \int_0^T qdt) \quad \text{with } q = q(t)$$

Differentiating with respect to q and setting the result equal to zero gives:

$$\frac{\delta V}{\delta q(t)} e^{-rt} = \lambda$$

If, for example, we put $t = 0$ and $t = t'$, we get a result that is exactly the same as the one presented earlier in this exposition, because $\delta V/\delta q = MP$. Explicitly, we obtain $MP(t') = MP(0)e^{rt'}$. Observe from the Lagrangian that λ is the shadow price of ore in the ground. If, over the time horizon t, $\overline{K}$ is greater than cumulative production, then λ is equal to zero, giving $MP = 0$, and so $p = MC$, which is the standard profit-maximizing condition from elementary microeconomic theory. λ can also be interpreted as a scarcity price or royalty, and is much larger than zero if the resource is in short supply. See Solow (1974) and Banks (1974). I would like to take the liberty of pointing out, especially to students of economics, that most of the recent work done on exhaustible resources, particularly in Norway, has no practical or deeper theoretical application. As I pointed out at the Tokyo meeting of the International Economic Association (1978), it seems extremely unlikely that any mathematical exposition of the basic concept could improve on Solow's paper—although Solow does leave unanswered some questions concerning expectations, and how expectations influence the stability of the optimal price path.

Backstop Technology

Now let us look at the exceedingly important concept of the "backstop" (or backstop technology) developed by Nordhaus (1974) in a brilliant article. A backstop is a known technology for producing a very large amount of a given resource at a known price. (Actually, Nordhaus thought in terms of an infinite amount.) Suppose, for instance, that coal was only used to produce electricity, but that there was (and is) a finite amount of coal. When, however, this coal runs out, we will still want electricity, and so it is necessary to think about an alternative technology for producing electricity. One candidate might be uranium and thorium burned in breeder reactors, because given the probable amount of uranium and thorium reserves available, they would be capable of producing electricity for a very long time. If, like this economist, the reader does not particularly like breeder reactors, then he or she can think in terms of fusion reactors.

Another example that would be useful here is oil from coal functioning as a backstop for conventional crude oil. At some price there is a technology capable of providing us with large (but not infinite) amounts of liquids that can fulfill the present functions of crude oil. Continuing, suppose that crude oil is finite in supply, with B units available that can be extracted for zero cost; and every year A units are extracted. Now, ignoring the concept of optimal depletion, this means that the oil will last $B/A = T$ years. At the end of T years, if we want to continue to use A units of oil every year, we must have a technology available that can produce this oil. Suppose we know that in T years a technology that will produce A units of synthetic oil per year, indefinitely, will cost Z dollars. Its cost today, at time t, is then $Ze^{-r(T-t)}$. In other words, if we put this amount in a bank, and the interest rate was r, then in $(T - t)$ years it would grow to Z, and we would be able to continue consuming oil.

Taking this argument farther, if we need A units per year, and the technology for producing A units costs Z dollars, the cost of obtaining one unit of oil in T years will be $Z/A = F(T)$. We thus have as the present price of a unit of oil:

$$p(t) = F(T)e^{-r(T-t)}$$

In other words, even though $MC = 0$, oil should have a price of $p(t) > 0$ if we intend to continue using oil after T years. Note also that the price of a unit of oil in T years—and on to the end of time—will not be F dollars, which is the cost of a nondepreciable machine capable of producing one unit of oil forever, but rF dollars, which is the interest cost on F. (There is no capital cost because, in this simple example there is no depreciation of the asset.) This can be verified immediately by noticing that:

$$\int_T^{\infty} rFe^{-r(t-T)}dt = F$$

One final observation: if more oil were found, the time when synthetic oil would be required would be pushed further into the future. Obviously, $dp(t)/dT < 0$. Also, since $T = B/A$:

$$\frac{dp(t)}{dB} = \frac{dp(t)}{dT}\frac{dT}{dB} = \frac{dp(t)}{dT} \cdot \left(\frac{1}{A}\right) \quad \text{so} \quad \frac{dp(t)}{dB} < 0$$

Similarly:

$$\frac{dp(t)}{dA} = \frac{dp(t)}{dT}\frac{dT}{dA} = \frac{dp(t)}{dT} \cdot \left(-\frac{B}{A^2}\right) \quad \text{so} \quad \frac{dp(t)}{dA} > 0$$

This last expression merely indicates that if A decreases, $p(t)$ falls.

4
The International Trade in Steam Coal

As pointed out earlier, the international trade in steam (or thermal or steaming) coal did not amount to much before the first oil price shock. After that event, when it was discerned that the age of oil might be drawing to an end, this trade began to escalate. Certain international organizations—such as the International Energy Agency (IEA)—presented coal as the ideal bridging fuel between crude oil and such things as synfuels and renewables, while the celebrated World Coal Study (WOCOL) called its elaborately produced final document, *Coal: Bridge to the Future.*

Given this background, it would be pleasant to say that this chapter will develop an efficient general equilibrium or econometric model for the forecast of international trade flows in steam coal, but instead I must begin this presentation by making the uncomfortable statement that no such model can be constructed at the present time, despite the fact that such models do exist and are used for forecasting, and the energy-related literature is unfortunately filled with references to them. For this reason such prominent studies as *Steam Coal: Prospects to 2000,* prepared by the IEA and the WOCOL study, probably represent a state-of-the-art approach to the noble art of economic forecasting, even though a number of the conclusions drawn by these studies are patently invalid. As is well known now, a large-scale expansion in the consumption of coal does not represent a viable option unless both the short- and long-run environmental problems associated with coal can be solved, and nobody now knows exactly how this can be done. This matter is taken up in considerable detail in chapter 6; however, the principal problem is the financing of expensive investments in pollution suppression equipment in a period in which there is a general decline in capital investment.

The Pattern of Trade

In 1980 the total world trade in hard coal (both steam and coking coal) amounted to almost 240 million metric tonnes (240 million tonnes = 240

Mt), while the total trade in steam coal came to 79.5 million tonnes. Table 4–1 shows international steam coal flows in 1980, and a similar construction for the total trade in hard coal will be given later in this chapter. These and other statistics show that during the present decade the trade in steam coal has averaged 30 to 35 percent of the total trade in hard coal. At the same time it is clear that this figure is going to change drastically in favor of steam coal—though not, perhaps, at the rate predicted a decade ago.

As already indicated, economists who feel that the flows in table 4–1 can be explained by employing the usual elementary theoretical assumptions are in for a serious disappointment. It has been suggested, and it may be true, that the world coal market is highly competitive, which means that, in theory, the price of coal would tend toward its long-run marginal cost. For our purposes, this cost can be estimated as the average cost of high-cost coal in the country or countries that can supply coal at the lowest cost. All the countries in the lefthand column of table 4–1 have some low-cost coal, but in practice a perfectly competitive market would mean that most or almost all of world coal would, in the long run, be delivered by South Africa and Australia, assuming that these countries were willing to deplete their very large supplies of low-cost exportable coal in phase with the requirements of buyers.

There is, however, no indication of this attitude on the part of these sellers at the present time, nor are there signs that it will appear in the near or distant future. As pointed out in chapter 3, the government of South Africa limits its total exports through a comprehensive system of export licenses. A possible reason for this might be that the South Africans are thinking in terms of long-run as opposed to short-run profit maximization, and believe that by restricting world supply that country will be able to collect large rents or profits on its present exports. Without an unlimited supply of low-cost coal satisfying world coal demand, a considerable amount of high-cost coal will have to appear on the market, and the world price of coal will therefore be determined by these high-cost materials. It is not easy to get exact figures, but at one point in 1980 mine mouth costs in South Africa averaged $14 per tonne, and rail and loading costs $8.50 per tonne. With an f.o.b. price of $55 per tonne at Richards Bay, the mine mouth profit was $32.50. In a similar vein the Australian federal government must approve export contracts, and it has also compelled producers to coordinate their negotiations with foreign buyers.

On the other side of the market, the Japanese government and others practice a highily sophisticated version of diversification, making sure that their economies are not overly reliant on a single supplier. Their purchasing plans leave room for a great deal of high-cost coal. In

Table 4–1
World Trade in Steam Coal, 1980
(millions of tonnes)

From/To	*EEC*	*Other Europe*	*Far East*	*Canada*	*Other*	*Total Exports*
United States	10.7	1.7	1.5	9.8	0.6	24.3
South Africa	19.2	3.0	0.3	0	1.3	25.5
Poland	11.0	1.0	0.0	0	1.0	13.0
Australia	3.3	0.0	3.6	0	1.9	8.8
USSR	2.0	0.5	0.2	0	0.3	3.0
Other	2.8	0.0	2.1	0	0.0	4.9
Total Imports	49.0	6.2	7.7	9.8	6.8	79.5

Source: United States Bureau of Mines; *Australian Coal Reports*, 1979, 1980, 1981.

addition, it should be understood that we find a number of prices in the coal market, since coal is sold both "spot" (that is, for immediate delivery) and on short- and long-term contracts. Accordingly, buyers will sometimes choose to pay a higher price for coal if it means security of supply over a number of years. Recently, Danish importers have indicated their willingness to pay $10 a ton or more premium for U.S. coal delivered under long-term contracts.

Phenomena of these types are, to my way of thinking, formally inconsistent with simple competitive market theory, where the law of one price is supposed to prevail. Let us also understand that uncertainty, in the sense in which it exists in the coal market, results in what economic theorists call suboptimal resource allocation. In these circumstances, the search for a comprehensive theory of the international coal trade is best not begun, because if such a theory were discovered, someone might make the mistake of believing it. At the same time, there is no harm in stressing that, everything else being equal, low-cost coal has a distinct competitive advantage over high-cost coal, although we do not know how much of an advantage.

The last point that will be made in association with this topic is that despite the large number of mines and, perhaps also, mining companies, there are distinct oligopolistic tendencies on both the supply and demand side of the world coal market. In South Africa, the Transvaal Coal Association (TCOA) determines production quotas for its members, assigns markets, and runs an export profits pool. This association is also the main owner of the Richards Bay export terminal. Concentration is less formal in Australia, but the large Australian mining houses, conglomerates, and foreign multinational companies have a decisive position in the determination of Australian coal output and exports. For example, CSR controls at least 10 percent of Australian exports, and is involved in 25 percent of current or planned steam coal projects. On the importing side, state companies coordinate coal imports in France, Japan, and Spain. In Germany, the quantity of imports are strictly regulated by the federal government, while in Italy, Belgium, Denmark, the Netherlands, and Sweden national electric companies or utilities coordinate the import of coal.

History, Some Price Theory, and Coal Contracts

According to the *Shell Briefing Service*, the first known shipments of coal took place in 1325, when a vessel from Pontoise in France delivered a load of corn to Newcastle upon Tyne in England, and brought back a cargo of *charboun de meer,* or sea coal. By the nineteenth century the United Kingdom was the dominant force in the world coal trade, aided by comparatively efficient mines, a good internal transportation system, a large merchant fleet, and the best facilities in the world for financing trade. In 1913 exports from the United Kingdom totaled 100 Mt, with much of this coal carried in ships of one or two thousand deadweight tonnes (dwt). "Deadweight" is the loaded displacement of a vessel minus the empty displacement, and therefore measures the cargo capacity of a ship. The United Kingdom retained its leading position until about 1925, when it was displaced by Poland and Germany. After World War II the United States, which did not begin to export coal until the end of the nineteenth century, began to develop its huge resources, and in 1957 U.S. coal exports peaked at 69 Mt of bituminous coal. Even then, however, the world trade in coal was declining in favor of oil, and a pronounced decrease was apparent until the beginning of the 1970s.

As a rule, only high quality coal enters international trade. Bituminous coal with a sulfur content of less than 1.5 percent is the usual specification. It can be easily shown, however, that in general there is no place for lower quality subbituminous coal in international trade. If θ_b is the heat content of high quality coal (in Btu/lb), and θ_s is the heat content of low quality coal, and with $\theta_b > \theta_s$, then for mines producing low and high quality coal, serving the same buyer, and incurring no transport cost, we must have the following relationship for mine mouth costs P_s and P_b:

$$\frac{P_s}{\theta_s} = \frac{P_b}{\theta_b}$$

which implies that:

$$P_s < P_b$$

With transport costs t_s and t_b the first expression becomes:

$$\frac{P_s + t_s}{\theta_s} = \frac{P_b + t_b}{\theta_b}$$

Now introduce a foreign market, with transport costs T_f from the original point of purchase. Thus we have $t_s \rightarrow (t_s + T_f)$, and $t_b \rightarrow (t_b + T_f)$; but this implies that we get:

$$\frac{P_s + t_s + T_f}{\theta_s} > \frac{P_b + t_b + T_f}{\theta_b}$$

since:

$$\frac{T_f}{\theta_s} > \frac{T_f}{\theta_b}$$

Clearly, a value of P_s $(=\overline{P}_s)$ can be found to bring this expression into equality, and so:

$$\overline{P}_s = \frac{\theta_s}{\theta_b}(P_b + t_b) + T_f\left(\frac{\theta_s}{\theta_b} - 1\right)$$

thus:

$$\frac{d\overline{P}_s}{dT_f} < 0 \quad \text{since } \frac{\theta_s}{\theta_b} < 1$$

With realistic values of θ_s/θ_b (for example, 9,000/12,000), this derivation indicates that the price of subbituminous coal would have to be well under that of bituminous to be competitive.

Spot and Long-Term Markets and Shipping Terms

A relatively small amount of coal is traded on the so-called spot markets, which can be thought of as markets for immediate delivery. The problem here is that hardly anyone knows exactly how much coal is involved, what immediate means, and in fact if there really is a spot market in, for example, Europe.

As far as I can tell, there is no possibility at the present time of determining just how much coal is traded on spot markets. Insofar as availability is concerned, I have heard 15 percent mentioned for the United States by a fairly reliable source, but the amount of spot coal traded is probably much less. On the other hand, it should be accepted that a spot market does exist in Europe. This is a problem that I have not considered in my previous work, but the issue is fairly simple. The presence of uncommitted coal stocks, and the fairly high expense of maintaining these stocks, will always encourage the formation of a spot market, though it is impossible to make any estimates of its size. The demand for spot coal is virtually assured by the inability of large consumers to judge future requirements exactly, and thus they will sometimes have a desperate need to make up deficits with spot purchases. Occasions will also exist when they will be equally anxious to get rid of excess supplies. In the United States stockpiles can often be found in the vicinity of coal mines, and in addition spot contracts can be filled by the ability of both large and small firms to increase production substantially in a relatively short time.

In the Benelux countries, large stockpiles from which spot sales can be made are quite common, and these inventories also supply the extensive Rhine barge trade in coal. At least a part of this barge trade is of a spot nature, since occasionally it is highly profitable to delay certain transfers of coal when important buyers who have underestimated their requirements suddenly appear. Similar activities take place on a much larger scale in the United States, particularly at Hampton Roads, where the practice of buying and exchanging carloads of coal among traders is highly developed. The expression immediate seems to refer, on average, to less than three months, although delivery times of up to one year are not unknown.

By way of contrast, long-term contracts can—in some isolated instances—stretch up to 20 or 30 years, particularly when it is a matter of arrangements between coal producers in the United States and electricity generating firms (that is, utilities). Unless I am mistaken, very little needs to be said on this particular subject at the present time, because without the possibility of making long-term agreements, the directors of coal mining companies would not be willing to make the huge investments required to mine and handle coal, nor could many large buyers—such as utilities—with unalterable commitments to their clients take the chance of buying coal on a spot market. These contracts, incidentally, can also be distinguished from contracts calling for more or less immediate delivery by referring to them as *forward* contracts—that is, contracts calling for physical delivery at some future date.

A point that is also relevant to the discussion in the introduction to this chapter about the degree of competitiveness of the coal market, is that it is inconceivable that a rule or algorithm will be derived which will indicate buyers' or sellers' preferences, at a given time, in the matter of contract durations, or the actual duration of contracts eventually negotiated between the partners in a deal. Thus, in theory and practice, it is possible to discern periods in which there are unchanging supply and demand conditions but different market prices because contract durations have changed due to changes in the bargaining power of buyers or sellers. At the same time, it should be clear that there are no fixed relationships between contract durations and prices. All this depends on the subjective evaluation of future supply and demand by traders, and their bargaining power, and it so happens that bargaining power is something that is generally not quantifiable. In some situations, the prices on very long-term contracts will be comparatively low relative to shorter contracts, and in others comparatively high. Developments have been moving in the direction of permitting ex post modification of the prices on long-term contracts, which does not mean just escalation clauses, but also provision for reexamining or renegotiating prices from time to time (generally from two to five years). The arguments for changing the conditions or terms on contracts generally turn on the impossibility or near-impossibility of buyers and sellers adjusting to unforeseen cost or exchange rate changes. The system does not exist in coal, as it does in oil, of pricing almost exclusively in dollars; but the importance of the United States in the world coal trade, as well as the interest held in many multinational coal companies (especially in Australia and South Africa) by U.S. firms, probably makes the dollar the most widely used currency in transactions involving coal. On some contracts, currency risks are partially eliminated by automatic escalation or deescalation provisions associated with either the negotiated price or with various costs. This practice is not yet widespread, but unless the growing instability of exchange rates can be checked, it will probably be applied more liberally in the future.

There is also no means of determining where or when a contract will be written f.o.b. (free on board, or delivered onto ship at the export terminal) as compared to c.i.f. (cost insurance freight, or delivered to some point in the importing country). Buyers of coal who feel that they can profit from arranging ocean transport prefer to buy f.o.b., just as sellers in the same situation prefer to sell c.i.f. At the present time most contracts seem to be written f.o.b., but this situation could change overnight and is quite unpredictable. It has been true for much of the last decade, however, that U.S. coal producers have done everything possible to avoid accepting any responsibility for the shipping of coal, because only the largest firms are in position to accept the risks associated with being forced to wait to load or unload. (These risks are called "demurrage" risks, and Demurrage on a 60,000-ton collier in Chesapeake Bay can amount to $15,000 per day). American producers tend to prefer their contracts written f.o.b. pier, or f.a.s. (free along side). In this way they avoid all contact with the maritime carrier.

The Transportation and Handling of Coal

An international coal chain begins at a coal mine and continues by rail or barge to a coal export terminal. Coal moves between the export and coal import terminal via the medium of dry bulk vessels or combination carriers capable of carrying both oil and dry bulk cargo; and from the import point it usually moves by rail or barge to the place where it will be consumed. These activities will be examined in some detail, because although it is not commonly realized, transportation (and handling) costs from mine to market will often comprise more than one-half of the total delivered cost of coal, and in some cases have amounted to two-thirds of this cost. On the average, some 30 to 50 percent of transportation costs can be imputed to inland transportation (rail and barge); 35 to 45 percent to sea transportation; and 20 to 25 percent to terminal costs. To get a rough idea of how much this can amount to in dollars, we can consider some statistics from the *Shell Briefing Service* for 1980 and 1981. On a worldwide basis, the cost of rail track, including signaling and auxiliaries, was $0.5 to $1.5 million per kilometer; a 100-car unit train cost between $7 million and $9 million; while a coal terminal (including berths but excluding other facilities) cost $12 to $15 per annual tonne of capacity. The cost of a ship of 75,000 dwt in that year was about $33 million, and for one of 150,000 dwt about $55 million.

Inland Transport

The inland transport of coal is usually handled by train or barge, although transport over short distances is possible by truck or conveyor belt. Another medium is the slurry pipeline, where coal moves in a slurry form (that is, coal and water), and for large volumes and fairly long distances, this technique is considered to be more economical than railroads. Slurry pipelines have been constructed in the United States, France, Poland, and the USSR. The slurry pipeline that transports 5 million tonnes of coal per year 450 kilometers from the Kayenta mine in Arizona to the 1580 MW Mohave power station in Nevada is often cited in the commercial literature. Some environmental problems are solved by using these pipelines, but they require a great deal of water, and environmental difficulties might be created because of having to dewater the slurry at the end of the pipeline. Other drawbacks of a slurry include the rapid erosion of pipelines and "derating", or power loss in boilers, caused by the water content of a slurry.

As pointed out in chapter 3, the most economical rail transportation system consists of unit trains operating on special routes and employing high-speed loading and discharging facilities. Unit trains vary in size, but according to the *Shell Briefing Service* consist of 50 to 139 cars with a total net load of 4,000 to 13,000 tonnes. Similarly, high-speed loading and discharging barges are common both in the United States (for use on such waterways as the Mississippi and Tennessee rivers) and on the inland waterways of Europe (for example, the Rhine). They also operate in the coastal waters of Japan.

As things have now developed, the cost of inland transport has become a key factor in making U.S. coal less competitive in foreign and U.S. markets. Rail rates in the United States add between $14 and $18 to the cost of coal, and in the domestic U.S. market some utilities have claimed that they occasionally pay more for the shipping of coal than for the coal itself. Furthermore, the U.S. railroads are actively trying to stop the expansion of slurry pipelines. A long-distance slurry pipeline, costing $3.8 billion and designed to carry 22.5 million tonnes of coal a year a thousand miles from Wyoming's Powder Basin to the U.S. Gulf States, has been cancelled because it has been outbid by a railroad. Inherently there is nothing wrong with this outcome, although by increasing the capability of railways to block the right of way of slurry pipelines over land that is owned by railways, the costs of slurry pipelines are raised above what might be regarded as a socially optimal level. At the same time the ruling of the Interstate Commerce Commission, which gives railroads their pricing freedom, theoretically makes it possible for these carriers to underbid competition from other transportation media in certain districts by virtue of being able to raise prices elsewhere.

Terminals and Harbors

Terminals are necessary to transfer coal from one type of carrier to another. Least-cost loading and discharging probably requires terminals that can accommodate ships of up to 150,000 tons, but on the loading side it will take years before the U.S. West Coast is adequately supplied with such installations. Furthermore, on the eastern and Gulf coasts, although plans exist for a broad upgrading of facilities, and a great deal of work has taken place during the past five years, harbors are still limited to colliers (that is, coal carriers) in the Panamax class, which means less than 70,000 dwt. Supercolliers cannot take on a full load because at 100,000 to 110,000 dwt they require a channel depth of 50 feet, as against the available 45 feet. It has been said that deepening the Baltimore channel to 50 feet could mean a decrease of about $4 to $5/tonne in the current cost of shipping coal to Northeast Europe, and it has been argued by both coal producers and terminal managers that provisions should exist to permit the deepening of channels to 60 feet. Unfortunately, however, such things as budget authorization and permit reviews can take up to twenty years, although the present U.S. government apparently sympathizes with the need to reduce this period drastically.

Loading and unloading costs, on the average, are between $1.5 and $2 per tonne over most of the world. Exporting countries with highly efficient port facilities are Australia and South Africa; however, by the

1990s it seems reasonable to expect that the resolution of its inland transportation problem will make it profitable to expand terminals in the United States to the optimal size. On the importing side, very large terminals either exist or are being built in Western Europe and Japan. Both Rotterdam and Hamburg can handle ships of 150,000 dwt, and in a very short time Rotterdam should be able to receive ships of 250,000 dwt. The situation in France and Italy is obscure just now, because there is some uncertainty as to the amount of coal they will eventually use, but LeHavre and Dunkirk—which already can receive all except the largest ships—can be expanded to take ships of up to 150,000 to 180,000 dwt, and the same is true of a number of ports in Italy (for example, Trieste and Milazzo). As to be expected, terminal capacity already exists or is being constructed in Japan to accommodate, at minimum cost, almost all planned increases in that country's consumption of both steam and coking coal. Australia supplied over half of the 76 million tonnes of coal imported by Japan in 1983 (17 Mt was produced domestically). Forecasts of Japanese demand have been falling steadily, and according to Japan's Ministry of Industry and Trade, the total demand for coal in 1990 will be between 107 and 113 million tonnes, with 45 to 48 million tonnes consisting of steam coal.

Modern export terminals for coal can load ships of 120,000 to 150,000 dwt in three or four days. In general, however, more time is required for unloading. At the major ports in Europe—for example, those mentioned previously—the average unloading rate is about 25,000 tonnes/day.

Maritime Transport

About 10 percent of worldwide coal shipments are arranged on the spot market; another 10 percent on a trip basis, with a ship chartered for a trip or a round trip; and the other 80 percent of voyages arranged on a longer term basis, which is sometimes called "time chartering" and involves a ship being chartered for a year or two by some coal buyer to cover his contractual purchases over that period, or the arranging of a contract of affreightment that covers anything from a few voyages with small cargoes to large volumes over a period of years. Once again, in dealing with long-term contracts, we realize that establishing the length of contracts is an art rather than a science, since the trick for the purchaser of shipping services is to sign the longest possible contract at a time when it appears that shipping prices are at bottom. Freight rates were considered exceptionally low two years ago, and the demand for long-term contracts was correspondingly high; but the persons who bought shipping space on long-term contracts at that time have since seen prices fall by 50 percent.

The price of shipping is basically determined by supply and demand, and today's low shipping prices can be traced to the inability of shipping owners and their bankers to make realistic estimates of the future demand for coal. According to the WOCOL study, the expected enormous increase in the world consumption of coal presaged an equally large expansion in the building of maritime coal carriers. WOCOL insisted that bulk tonnage requirements had to grow from approximately 36 million dwt in 1975, which happens to have been the maximum output of the entire world shipbuilding industry at that time, to 50 million dwt at some time between 1985 and 1990. But at the beginning of 1984, 6 percent of the world bulk carrier fleet was laid up, while at the same time large numbers of vessels were scheduled for delivery. In these circumstances, even a recovery of the entire world economy comparable to that which took place in the United States during 1983 and 1984 would not cause the ordering of more ships, although if it were true (which it is not) that the world coal trade will or might expand the way WOCOL and the IEA thought it would (making possible a Western European consumption of 416 Mtce/year by 1990, according to WOCOL), then a huge demand for ships would develop. Table 4–2 provides information on the price of shipping coal.

Although it may not be immediately evident, falling freight rates due to changes in demand (or, as will be discussed, economies of scale) tend to make coal that is carried longer distances more competitive if its mine mouth or f.o.b. cost is comparatively low (for example, Australian coal in Europe or U.S. and South African coal in Japan). This can be illustrated with some very elementary algebra.

Begin with a situation where the delivered price of coal is higher if purchased from country B than from country A, with the delivered price of coal defined as the f.o.b. price (F) plus the transport cost (T). The statement of the problem gives:

$$F_B + T_B > F_A + T_A$$

with respect to a third market. Now assume this is so because $T_B > T_A$, although $F_A > F_B$. In the real world, this arrangement would correspond to the situation for Australia (B) and the United States (A) with respect to the West European market three or four years ago. Next, suppose that there is a uniform reduction in transport costs for each country of x percent. Costs are now:

$$F_A + (T_A - xT_A) = F_A + T_A(1 - x)$$
$$F_B + (T_B - xT_B) = F_B + T_B(1 - x)$$

Table 4–2
Typical Freight Rates, 1981–1984
(dollars/tonne)

	January 1984	*January 1982*	*January 1983*	*December 1983*	*February 1984*
Hampton Roads–Japan (50,000 dwt) (via Panama)	28.5	17.75	13.25	12.5	—
Hampton Roads–Japan (100,000 dwt) (via Panama and South Africa[1])	25.50	13.50	10.75	10.5	11.50
Australia–Japan (unweighted average)		9.00			7.20
Hampton Roads–NW Europe (50,000 dwt)	17.00	7.50	6.50	5.5	—
Hampton Roads–NW Europe (80,000 dwt)	—	—	—	4.75	5.45
South Africa–Northwest Europe	—	11.0	—	—	7.40
South Africa–Northwest Europe	—	10.0	—	—	—

Source: *Coal Week* (various issues).
[1]Topping off at Richards Bay.

The question we will now ask is what value of x is necessary to reduce the delivered price of B's coal to a third market to a level lower than that of A's coal. Put symbolically, this means:

$$F_A + T_A(1 - x) > F_B + T_B(1 - x)$$

or

$$(T_B - T_A)x > (F_B - F_A) + (T_B - T_A)$$

This can be rewritten to give:

$$x > 1 + \frac{F_B - F_A}{T_B - T_A} = 1 - \frac{F_A - F_B}{T_B - T_A}$$

To see how this functions, we can use some approximate values for Australia and the United States during a short period of three years ago: $F_A = 58$, $T_A = 15$, $F_B = 52$, and $T_B = 27$. From the above inequality we get:

$$x > 1 - \frac{58 - 52}{27 - 15} = 1 - \frac{6}{12} = 0.5$$

In other words, if the ocean shipping cost fell by more than 50 percent, Australian coal would be cheaper in Europe than coal from the United States. Apparently this has taken place over the past three years. On this point it might be instructive to consider some average export costs for a short period in 1981. The f.o.b. price shown in table 4–3 was the price on new contracts.

Because of economies of scale, coal is carried in the largest vessels compatible with port and storage facilities, except in those cases where it is more economic to use a smaller ship to fit the requirements of the backhaul trade of another commodity. As an example, an average saving of 25 to 30 percent is possible with a ship of 150,000 dwt instead of 60,000 dwt—assuming, of course, that there are buyers to the cargos of these ships; and further, though incrementally smaller, savings are possible up to 200,000 to 250,000 dwt. One of the main problems with the larger ships is that they cannot go through the Panama and Suez canals, but plans exist for a deepening of the Suez Canal so that it can accommodate vessels of up to 150,000 dwt. It is also considered likely that a gradual reintroduction of the coal-fired ship will take place before the end of this century, particularly if the oil prices begin to escalate

Table 4–3
Typical Export Costs in Panamax-Type Vessels (60,000–80,000 dwt), 1981

	Price f.o.b.[a]	*Shipping*[a]	*Unloading*[a]	*Delivered Price*[a]	*Dollars M/Btu*
To Europe					
South Africa	41	20	2	63	2.90
United States	55	15	2	72	2.85
Australia	50	27	2	79	3.40
Poland	46	5	2	53	2.20
To Japan					
Australia	50	18	2	70	3.00
Canada	49	15	2	66	3.30
South Africa	41	22	2	65	3.00
United States	56	25	2	83	3.20

[a]dollars/tonne

again. Although ships have been exclusively oil-fired during the past twenty-five years, 20 percent of the world's shipping was coal-fired in the late 1940s. The effect of the influence of vessel size and bunker (that is, fuel for ships) costs can be seen in figure 4–1.

Backhauling can sometimes offer the chance to make an appreciable savings in transport costs. For instance, the ships that take coal from Australia to Europe can, on the way back from Australia, move iron ore or bauxite from Brazil or Africa to markets in the Far East. In 1979, according to WOCOL, an average saving of $6.3 per tonne was possible for a 150,000 dwt vessel carrying coal from Australia to Northern Europe, and returning with a backhaul.

The Soviet Union and Australia

By way of completion, a few more comments are required about the coal sectors of the USSR and, especially, Australia. Taken at face value, present statistics suggest that the Soviet Union will eventually become the Saudi Arabia of coal. As it happens, though, the Soviet coal fields

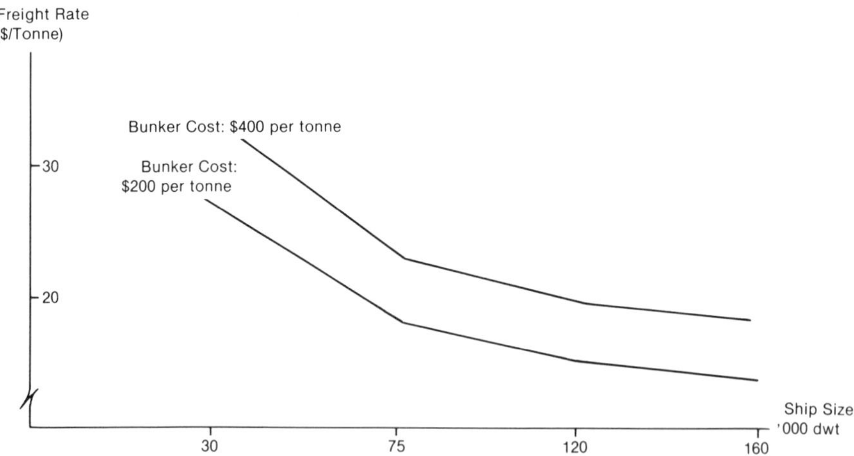

Source: IEA and Kjellman (1981).

Figure 4–1. Freight Rates as a Function of Ship Size and Bunker Costs

that are easily accessible to export markets are severely depleted. In the Donetsk Basin, where almost 30 percent of Soviet coal is produced, mining now takes place almost a kilometer under ground, and seams are moving deeper; they may be at 1,500 to 1,600 meters by 1990. Similarly, the Soviet coal fields in Asia and the Arctic are underdeveloped to a point where even billions of dollars in new investment would not guarantee a major increase in the availability of steam coal to Soviet industrial centers and export markets. The coal mining areas of eastern Siberia are under permafrost. Still, it should be appreciated that of seven giant coal basins (of a global total of 2,100 major deposits), five are in the USSR.

As is the case in other countries, the Soviet government hopes to get a proportionally larger amount of coal from open pit mines in the future, and official calculations suggest that there are huge productivity and cost gains to be realized. The question is, though, just how long in the future will it be before the authorities have introduced the technology that will allow these gains to actually appear. The development of suitable draglines and mechanical shovels has been much slower than anticipated, and Soviet dump trucks have been too small. At the Neryungrii mine, which is exporting coal to Japan, U.S. and Canadian vehicles had to be employed.

If the new open cast mines can meet expectations, much more coal will be introduced into the Soviet electricity generating sector in place of oil. Very large open pit projects are contemplated for Ekibastuz in northern Kazakhstan, which will probably be combined with a major electricity generating development. The same is true for the Kansk-Achinsk lignite basin in southern Siberia, where low-grade coal will be used in power stations directly adjacent to the coal deposit. The electricity that is produced will be transmitted by very high voltage lines to industrial cities in European Russia.

An important advance in the Soviet coal sector has been the fairly high degree of wage differentiation in favor of miners working underground or in harsh conditions. Some of these miners work only 30 hours per week. Unless I am mistaken, coal lies fairly low on the Soviet list of priorities, but in the long run it will be fairly difficult for the decision-makers in that country not to devote the same attention to coal that they have given to gas. It should be noted that after a ten-year construction period, the Baikal–Amur Railway running from central East Siberia to the Pacific Ocean is nearly complete. This railway will be a key facility in the exploitation of East Siberia's enormous hydrocarbon wealth, and an important step in the successful eastward displacement of the USSR's energy base.

Australian Coal

Not long after the colonization of Australia, coal was discovered near a settlement that is now called Newcastle, which has become one of the most important coal mining centers in Australia. Several years later, in 1979, a coal outcrop was sighted near Coalcliff, south of Sydney and in the vicinity of Wollongong, which is also one of the most important coal mining localities in that country; the following month Lieutenant Shortland located a coal deposit at the mouth of the Hunter River. It has now been officially decided that the first coal mined in Australia came from outcrops on a cliff face of the Hunter River estuary. This was in 1799.

In 1801 about 150 tonnes of Hunter River coal was exported to India, and later the same year coal was shipped to the Cape of Good Hope. The coal mining district in Australia gradually began to spread, and mines worked by convict labor were soon established at Newcastle and elsewhere in the vicinity of Sydney. As for Queensland, coal was first located near Ipswitch in 1827, but not until 1870 and the discovery of the Aberdare seam did the Queensland coal mining industry became important.

There is also coal in the states of Victoria, South Australia, Tasmania, and Western Australia—although the deposits in these latter three states are relatively unimportant, and the brown coal of Victoria is not exported. The main coal-producing *basins* or deposits are Sydney Basin; Bowen Basin (Queensland); Clarence-Moreton Basin (located largely in Queensland, but stretching into New South Wales); Galilee Basin (Queensland); and the brown coal deposits of the Latrobe Valley in Victoria.

The growth of Australian production was hindered for a long time by the resistance of mine workers to a comprehensive mechanization of the mines. The big issue was fear of unemployment, although the primitive machinery employed during the nineteenth century also jeopardized the safety of the miners. When mechanization did arrive in Australia, it arrived with a rush; today, output per man-shift is among the highest in the world—some say the highest. Another factor contributing to the jump in productivity in Australia was the shift to open pit (or open cut) mining that began in the 1930s. Eventually the coal mining sector was able to become extremely capital intensive and, as a result of the nature of open pit mining, safer rather than less safe. Tremendous efforts were also made to avoid the environmental damage associated with open pit mining. For instance, Queensland and New South Wales companies are required to lodge a bond of $123 (Australian) per hectare as a pledge against restoring land worth $2 per hectare.

Both Australian mine owners and the Australian government would like to see a much larger export of Australian coal. In fact, if the demand is forthcoming, the intention is to raise coal exports by a factor of four or five in the coming twenty years. In 1978–1979 Australian coal exports came to more than 38 million tons, which represented 53 percent of Australian's salable output. Japan bought 66 percent of this amount; Europe received 23 percent, Korea 5 percent, and Taiwan 3.5 percent. Plans exist for exporting 200 million tonnes by the year 2000 if the market exists and if Australia possesses the port and transport infrastructure needed to meet a growing world demand. At present the coal mining industry is one of the largest contributors to government revenue of any industrial sector, and every dollar earned from the sale of coal results in tax revenues of 36 cents. Of this amount 21 cents went to state governments as royalties (since resources in the ground normally belong to the local community—that is, the state), land tax, payroll tax, shipping fees, rail freight, and other service charges. It is interesting to note that tentative plans have been drawn up for the large-scale use of pipelines to transport coal to Australian markets or overseas terminals.

A major problem that will have to be met with a rapid expansion of the coal industry is the growing cost of both production facilities and infrastructure. The cost of infrastructure has risen so rapidly that it has become more profitable to exploit coal deposits by underground mining in or near the built-up areas of New South Wales, than to develop open cut mines in some of the more remote areas of the country (assuming there is no difference in the richness of the coal deposit). In New South Wales the development of an underground mine designed to produce 1 million tons of coal per annum can be expected to cost $25–$100 million and to take from three to five years. By the same token, to develop an open cut mine producing 3 million tonnes of coal per year in Central Queensland will normally cost about $400 million, but could cost much more. For instance, a 45-cubic meter dragline used to remove soil and rock lying above the coal seam can cost as much as $20 million.

The Australian coal industry is privately owned. The big names in this industry are Utah Development Company, CRA, and Shell Australia. A considerable amount of the coal industry is foreign owned or controlled, and in percentage terms foreign owners may control more than for the Australian minerals industry as a whole (where the figure is about 52 percent, with firms from the United States and United Kingdom holding about 80 percent of this). There is a considerable amount of antagonism in Australia toward the foreign ownership of mineral resources, and therefore it seems unlikely that foreigners will be encouraged to acquire more mineral properties in that country. At the same time, expectations are that the Australian energy sector will expand more than any other sector; and as things now stand this is patently impossible without more overseas investment—and presumably more overseas ownership, at least in this sector.

Regardless of the ownership issue, however, it is generally recognized throughout the country that the mining industry has made an important contribution to the welfare of the average Australian, particularly for those individuals who are employed in this industry. Wages in coal mining are much higher than the average Australian wage; and the total tax burden (which in the coal industry is augmented by certain levies) is higher for producers than in the United States, Canada, and South Africa. (The Australian company income tax is 40 percent of taxable income, with taxable income defined as sales revenue minus operating costs, interest charges, depreciation allowances, and special deductions such as investment allowances.) It could be reasoned, however, that the larger tax burden (of which a reasonable part falls on foreigners) is a payment for the kind of political stability that does not always exist in other parts of the mineral-producing world.

If we consider the experiences of the largest importer of Australian coal—Japan—it is difficult to conclude that the Japanese have come to regret their association with the Australian coal industry. In the past Japan has mostly purchased coking coal from Australia, but in the coming years the trade in steaming coal will undoubtedly be more important—as is even more the case for European importers of Australian coal. Coal exports to Japan have generally taken place under long-term contracts having a duration of up to fifteen years; but although quantities are specified, plus-minus options of as much as 10 percent are allowable, subject to prior notification. At one time prices were set in terms of a base price that could be escalated for certain costs; but present arrangements often permit a price adjustment that is a function of market conditions. Coking coal contracts with Japan generally carry one of the following stipulations:

1. Regular price reviews can take place outside contract provisions.
2. Contracts provide for regular reviews, although price is expressed in terms of a base price and escalatable components, with definite limits being placed on these escalations.
3. Contracts may provide for the price to be set by annual or biannual reviews, with a no-price, no-contract clause. Obviously, this type of arrangement is virtually equivalent to a short-term contract.

Because of some bad feeling that has surfaced in Australia concerning the alleged abuse of coal and iron ore contracts by their overseas customers, it has been suggested that new contracts contain the following features:

1. Pricing reviews should occur at intervals of not less than two years.
2. If buyers and sellers cannot agree on a pricing formula, then it should be the option of sellers to determine whether the contract should be terminated with some notice or whether the existing pricing arrangements should continue.
3. There should be a limited scope for adjustment of contract tonnages.
4. If possible, prices should be denominated in a basket of currencies to take account of currency instability.
5. Most contracts today are on an f.o.b. basis. More contracts on a c.i.f. basis should be resorted to.

The trade matrix shown in table 4–4 makes clear the position of Australia in the world trade picture. The consumption of coal in Australia has increased during the postwar period by 2.5 percent a year on the average, although in the last decade the rate of increase has been about 3 percent. Where exports are concerned, the intention is to sell as much coal as possible, subject to the constraints mentioned earlier.

The brown coal of Victoria is another story. Since its heating value is only one-quarter to one-third that of black coal, there is no possibility of its being exported. For many years it appeared that brown coal would not even have a market in Australia, since it could not compete with domestic gas. In a world of rising energy prices, however, the situation has changed; among other things, brown coal has become very important for the industries of Victoria—in particular aluminum refining. The thickest seam of brown coal in the world is found at Loy Yang, and the electricity generated from this coal may be some of the most inexpensive in the world. Some Australian economists have argued that this cheap electricity should not be used to subsidize the Australian aluminum industry.

Here it should be observed that Australia has a very high unemployment rate, with many of the unemployed receiving subsidies in the form of a dole. By transfering a small amount of these subsidies over—for example, to the aluminum-producing and using sector—it might eventually be possible to give a very large number of these people work, particularly if there are increasing returns to aluminum refining (which is likely) or the fabrication of products using aluminum. In fact we know, or should know, from theoretical welfare economics that if persons are unemployed because of the absence of an input, that input is worth the whole value of the product that has been lost, and not merely the

Table 4–4
International Hard-Coal Trade Matrix, 1979
(millions of tonnes)

From/To	*Benelux*	*France*	*Germany (FR)*	*Italy*	*U.K.*	*Canada*	*Japan*	*CMEA*[1]	*Others*	*Total*
European communities	5	8	1	2	–	–	1	–	1	18
United States	4	3	2	4	1	17	13	–	12	58
Canada	–	–	1	–	–	–	11	1	1	15
Australia	1	2	1	1	2	–	27	4	1	41
South Africa	2	8	1	2	–	–	3	–	5	23
USSR	–	1	–	1	–	–	2	15	3	23
Poland	2	4	2	3	1	–	–	15	6	34
Others	1	1	1	–	–	–	1	2	3	10
Total	16	29	9	15	6	17	60	38	32	222

Source: Adapted by permission of the publisher from Ferdinand E. Banks, *Resources and Energy: An Economic Analysis* (Lexington, Mass: Lexington Books, 1983).

Note: Many values have been rounded; (–) signifies less than 1 Mt.

[1]Centrally planned countries of eastern Europe (excluding the USSR).

amount that would have been paid for it in some market. In these circumstances, the actual value of Australian brown coal (and perhaps also black coal and gas) is higher than the price it would bring in domestic or foreign markets.

To see how true this is, we have only to observe the nascent New Zealand aluminum refining and fabrication industry, which has apparently become a success although operating under much less favorable conditions than would be the case for similar Australian facilities. The New Zealand refinery came on stream in the middle of the recent world recession, but by operating at full capacity it generated the scale economies that made it fully competitive both domestically and internationally. Now, in the wake of rising aluminum prices, new investments are being contemplated, particularly in the further processing of aluminum. At the present time aluminum wheels fabricated in New Zealand are being exported to Australia, and given the amount of aluminum metal being produced, as well as the small amount of this metal being absorbed by the New Zealand economy, additional processing into fabricated items is probably warranted strictly on the basis of export diversification.

Despite testimony given a few years ago to an Australian Senate committee about the absence of spin-off or multiplier effects that result from the smelting of aluminum, thousands of new jobs were created in and around Invercargill, New Zealand, as the direct result of the establishment of a large aluminum smelter, although many other industrial projects started at the same time did not do well.

Appendix 4A
Interdependent Activities

The problem that will be investigated here, more or less superficially, was raised in the last part of chapter 4 in connection with the valuation of Australian energy resources. It was claimed that an important, though apparently little known, result from welfare economics is that when persons (or other factors of production) are unemployed because of the absence of an input, that input is worth the whole value of the product that has been lost, and not merely the amount that would be paid for it in some market. This remark, incidentally, follows almost directly from Kuhn-Tucker theory.

A straightforward method of looking at this issue is via a simple parametric programming exercise in which a commodity x_2 is produced using a primary input R and a produced input x_1. This produced input—for example, gas or coal—uses only R in its production, and for our purposes R could be labor or, for that matter, a composite input that features labor and capital in a fixed proportion. Our production functions here are, of course, $x_1 = x_1(R)$ and $x_2 = x_2(R, x_2)$, and these production functions are sufficient to illustrate the interdependency of these two production activities. This matter will be clarified later in the exposition. Next, assume that the price of x_1 on, for example, the world market is given by p_1''; but the price of x_1 to industry x_2 can be set at some other value, and this price will be called p_1'. Naturally, p_1' can equal p_1''. We can now write a linear program containing these ingredients:

$$Z = p_1''(x_1 - bx_2) + x_2(p_2 - p_1'b) = \text{Maximum} \qquad \text{(a)}$$
$$a_{11}x_1 + a_{12}x_2 \leqslant R \qquad \text{(b)}$$
$$x_1 - bx_2 \geqslant 0 \qquad \text{(c)}$$
$$x_1, x_2 \geqslant 0$$

Let us examine this program. a_{11} is the amount of R needed to produce one unit of x_1, while a_{12} is the amount required to produce one unit of x_2. But, in addition, b units of x_1 are needed to produce one unit of x_2. Note that $x_1 - bx_2 \geqslant 0$ implies that at least enough x_1 must be produced to make possible the production of x_2. Also, it does not make any sense in this exercise to have $x_1 = 0$, although it might be an economically sound move not to produce any x_2, and use all of the resource R to produce x_1 for sale on the world market.

Observe that in the objective function (a), the price p_1'' is not just multiplied by the total output of x_1, but by the amount that is sold on the world market—that is, the total amount produced minus the amount used as an input in the production of x_2. In other words, x_1 is produced, but $(x_1 - bx_2)$ goes on the market. Similarly, $(p_2 - p_1'b)$ is a kind of net price. It corrects the market price for the cost of x_1 that goes into producing x_2. By rewriting the objective function and obtaining $Z = (p_2 - p_1'b - p_1''b)x_2 + p_1''x_1$, it becomes clear that increasing p_2 relative to p_1'' will, other things being equal, increase the attractiveness of producing x_2 relative to x_1.

Figure 4A–1 shows two solutions to this problem. In the solution depicted by $\overline{Z}$, p_2 is high enough and p_1' low enough to make the production of x_2 profitable. The exact solution is at point A, where $\overline{x}_1$ of x_1 is also produced, but it should be noted that all the x_1 produced is used as an input to x_2. The reason I have said p_1' low enough instead of p_1'', is that p_1'' is given and p_1' is being set. At the same time we understand that if p_2 is high enough, p_1' can be set equal to p_1'' and our solution will still be at A.

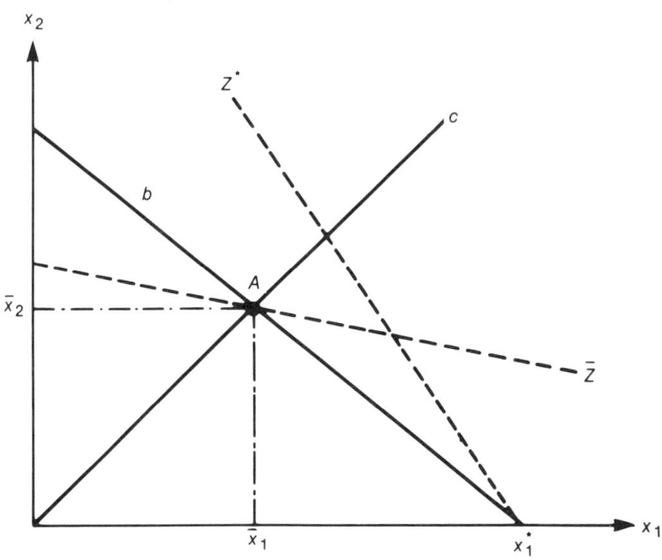

Figure 4A–1. Convex Polygon for a Parametric Programming Problem

By way of contrast, Z^* shows an arrangement where p_2 is so low, or p_1'' so high, that there is no feasible value for p_1' that will make the production of x_2 profitable. The solution of the program in this case is thus $x_1 = x_1^*$, $x_2 = 0$. By way of summation, the parametric nature of this exercise calls for choosing a value of p_1' that will maximize Z, given values of p_1'' and p_2. In line with earlier observations, this value could be under the world price. As indicated in the text, if there are nonlinearities and, in particular, increasing returns to scale in the production of x_2, pronounced departures from optimality might result if x_2 is not produced.

A more general approach to this topic features the maximization of a preference function $S(x_1, x_2)$ employing the production functions and resource availabilities as constraints. Thus we have:

$$S^* = S(x_1, x_2) + \lambda_1[x_1 - x_1(R_1)] + \lambda_2[x_2 - x_2(R_2, x_1)] + \lambda_3(R_1 + R_2 - \overline{R})$$

Differentiating this expression with respect to x_1, x_2, R_1, R_2, and setting the results equal to zero yields:

$$S_1 + \lambda_1 - \lambda_2 \frac{\delta x_2}{\delta x_1} = 0$$

$$S_2 - \lambda_2 = 0$$

$$-\lambda_1 \frac{\delta x_1}{\delta R_1} + \lambda_3 = 0$$

$$-\lambda_2 \frac{\delta x_2}{\delta R_2} + \lambda_3 = 0$$

These expressions can be easily rearranged to give:

$$\frac{S_1 - \lambda_2 \dfrac{\delta x_2}{\delta x_1}}{S_2} = -\frac{\lambda_1}{\lambda_2} = -\frac{\delta x_2/\delta R_2}{\delta x_1/\delta R_1} = \frac{\delta x_2}{\delta x_1}$$

In obtaining $\delta x_2/\delta x_1$, remember that $\overline{R} = R_1 + R_2$, and so $dR_1 = -dR_2$. If we did not have $x_2 = x_2(R_2, x_2)$, we would have $\delta x_2/\delta x_1 = 0$, and so:

$$\frac{S_1}{S_2} = \frac{\delta x_2}{\delta x_1}$$

However, we do have $x_2 = x_2(R_2, x_2)$, and so the optimum no longer corresponds to a perfect competition market equilibrium. Without going any further into the matter let me point out that this solution corresponds to two firms maximizing the sum of their profits, instead of each attempting to maximize profits separately. Unfortunately the market will not compel the observance of this arrangement, although a merger of these two firms (or industries) would tend to bring about what Malinvaud (1969) terms a market equilibrium.

5
Coal Topics and the Economics of Natural Gas

In this chapter a number of supplementary topics will be taken up, though only briefly. They include synthetic oil and gas (that is, synfuels), underground coal gasification, and petrochemicals from coal. The principal discussion will be of natural gas, since I want to use this opportunity to extend the observations I made in my book, *The Political Economy of Oil,* and to emphasize my belief that future energy scenarios should pay more attention to natural gas. There is also the matter of Melvin Conant's important observation in the book edited by Paul Tempest (1983): "The full development of Gulf gas—assuming that the terms for gas supply are competitive—will be the single most important consideration in world energy before the end of this century."

Coal Gasification

Techniques for the gasification of coal are almost two hundred years old. The first coal gas company that distributed its product for lighting was chartered in London, in 1812, while a similar company was established in the United States (in Baltimore) four years later. From about the middle of the nineteenth century until World War II, the technology for coal gasification was greatly improved in the United States. Eventually, however, manufactured gas in the United States was supplanted by inexpensive natural gas transported in pipelines. Further improvements in coal gasification then took place in Europe, particularly in Germany, where comparatively large amounts of coal were available. But here also natural gas came into the picture with the discovery of the rich Groningen (Holland) and North African fields; and in addition the growing availability of enormous quantities of inexpensive oil meant that only relatively small amounts of gas would be used anyway.

But after the first oil price shock, the decisionmakers came to the belated conclusion that there were inflexible limits on the supply of natural gas and conventional crude oil, and concluded that known coal gasification techniques should be upgraded to eventually produce large amounts of gas for use as a fuel and as a chemical feedstock. The technical details of coal gasification are not relevant to this survey, although several things are important. That is, although most coal gasification processes are similar, in that they involve the reaction of coal with steam to form carbon monoxide and hydrogen, the best known processes are the Lurgi process and the Koppers-Totzek (KT) process. The Lurgi process features sized coal descending into a gasifier where it is first dried and then carbonized by reacting with oxygen and steam. This is the process that is used in the Republic of South Africa to produce gas for use as a fuel and also synthesis gas for transformation into various chemicals and synthetic oil. As for the KT process, which seems to be gaining in popularity, steam and oxygen are used in the gasification step, and it will work with any type of coal. Three grades of gas can also be distinguished, depending upon whether air or oxygen is used for the carbon-oxygen reaction, or whether there is a methanation step that raises the heating value of the gas.

The first type is a low heat value and environmentally friendly gas that is of considerable value for power generation and industrial applications, but that must be used at its source because it is not economical to transmit this gas by pipeline. By environmentally friendly, it is meant that it is relatively pollution-free despite the use of high sulfur coal in its production. It may also be true that coal-gas plants emit only a tenth of the air pollutants associated with coal-fired electric plants of the same size. Next, there is a nitrogen-free and medium calorific gas that can be transmitted by pipeline over distances that are not too long and used directly for industrial and domestic purposes. This gas has a high value as a chemical feedstock, and as a source of hydrogen, methane, and methanol. Finally, there is a high calorific gas that, in one variant, is called substitute natural gas (SNG). SNG can supplement natural gas supplies in existing pipeline systems and is suitable for use with existing appliances.

In the United States, as a result of President Carter's Energy Security Act, a great deal of money was made available for experimenting with coal gasification and building pilot plants that were supposed to be the forerunners of large-scale commercial installations. As far as I can tell, the Lurgi process is being pushed hardest in that country, and several years ago it was estimated that by the year 2000, the equivalent of 1.6 Mbbl/d of oil would be produced; but in the last few years some momentum seems to have been lost in this endeavor. Among other things, the gradual deregulation of the natural gas price has caused many of the energy companies that were interested in coal gasification to increase their efforts to find more domestic natural gas or even to import liquefied natural gas (LNG).

Germany, however, with decades of successful experience in syngas experimentation, and large domestic supplies of coal, has renewed its effort to produce gas from coal on a very large scale. Rheinbraun has a scheme for converting 2 Mt/year of lignite into 1 Gcm/yr of synthesis gas (carbon monoxide and hydrogen). Full production will call for manufacturing around 350,000 tonnes a year of methanol, which will replace 500,000 tonnes/year of oil. Ruhrkohle has also solicited subsidies from the German government for a plant at Oberhausen in the Ruhr where 250,000/t year of hard coal would be transformed into 400 million cubic meters of synthesis gas. This gas would be used as a chemical feedstock.

Similarly, the chairman of British Gas has claimed that Britain now possesses the best process in the world for the manufacture of substitute natural gas from coal. This process builds on—but is not exactly the same as—the Lurgi system but is generally considered more versatile. It is called the British Gas-Lurgi Slagging Gasifier, and a refinement called the Gas Composite Gasifier is undergoing development in the United States. The intention in Britain is to develop a range of processes that will convert any hydrocarbon into a gas capable of replacing existing natural gas, without the objection of customers. This gas would first be added to the grid to help meet the peak demand in winter and its use gradually expanded. It has been estimated that fairly early in the next century coal gasification in Britain will require between 50 Mt and 150 Mt of coal per year. This should be compared to the 100 Mt/year that was the average output of the National Coal Board over the past two years.

Before finishing this section, let me note that large-scale coal gasification is still a rarity, despite the many successful pilot projects that have been developed. According to a 1982 estimate of the IEA's Research and Development Committee, coal gasification will not start to become economic in an OECD country before 1985–1986; and many member countries will have to wait until 1995 or later before they will be able to produce large amounts of these gases at an acceptable cost.

Oil from Coal

A dispute has been going on for some time now as to whether converting coal to oil is a more sensible activity than converting coal to gas. The IEA Research and Development Committee has claimed that with maximum investment coal liquefaction could become competitive by early 1990, and as a result it would be possible to substitute coal-based liquids for oil faster than synthetic gas could take over from natural gas. The problem here, of course, is that instead of maximum the term *optimum* should be used. But if we use optimum it means that we must be able, at the same time, to say something about prices and costs in 1990. If we cannot do this, economic theory tells us that this controversy is not very useful, and serious persons should not be distracted by it. As far as I know, prices and costs—particularly of conventional crude oil and natural gas—were not very prominent in this committee's deliberations, and as a result the quality of its conclusions should not be taken for granted.

There are three basic approaches to the liquefaction of coal: pyrolysis, hydrogenation, and gasification followed by conversion of the synthesis gas to liquids using Fisher-Tropsch or other technologies. Pyrolysis involves the destructive distillation of coal at a high temperature in the absence of air. The major product of this scheme is a char that can be used directly as a fuel, and also as a feedstock for fuel gas, synthesis gas, and a high grade of fuel oil. However, if the char contains a great deal of sulfur because high sulfur coal has been used, a hydrogenation step will be necessary.

The hydrogenation of coal is a process that has been known since before World War I. Liquid fuels can be produced from coal by direct hydrogenation; and at the same time it is recognized that hydrogenation under pressure is useful for improving the quality of some fuels while oil is being refined. During World War II, a number of hydrogenation plants operating in Germany produced about 4 million tons of oil and oil products per year. Another million tons were produced annually from the third gasification technique, which is sometimes called the Fisher-Tropsch synthesis. This is a slightly revised version of the process that is now being used at the Sasolburg installation in South Africa.

Sasolburg is constantly being expanded, and at present contains two main installations. Sasolburg I produces 240,000 tons/year of oil and Sasolburg II approximately 1.45 Mt/year. Production costs seem to be between $17 and $20 per barrel, of which half or a little more than half is the mine mouth cost of coal. Productivity at Sasolburg II is said to be 1.78 barrels of synthetic oil from every ton of coal. The reader should compare this figure with the heating value of a short ton of coal, which, in terms of conventional crude oil, is 4.35 barrels.

A great deal of effort has and is being made to adopt the Sasolburg process for large-scale use in the United States, and this effort falls to a certain extent within the sphere of interest of a governmental establishment known as the Synfuels Corporation, which originally was given the authority to spend $88 billion on research and pilot installations having to do with synfuels. The problem here though is that the sense of urgency that exists in South Africa—due to the partial oil embargo—does not exist in the United States. What originally existed, however, was an enormous bureaucratic jungle that potential synfuel entrepreneurs had to navigate, and some strange ideas on the part of the U.S. Congress, and perhaps the U.S. Department of Energy, that "small is good" and big oil bad. What this has meant is that the synfuels program in the United States has been a failure from the point of view of Congress, which originally directed the Synfuels Corporation to produce 2 Mbbl/d of synthetic materials by the year 2000, but a success for dozens—or perhaps hundreds—of relatively small firms that received hundreds of millions of dollars to construct uneconomic, demonstration-sized installations. To be fair, however, several of the larger pilot plants indicated a great deal of promise for the future. Perhaps the most successful was the Cattlesburg, Kentucky, plant financed by the U.S. Department of Energy, the State of Kentucky, Ashland Oil, Conoco, Standard of Indiana, Mobil, and the Electric Power Research Institute, using the H-coal process developed by Dynalectron Corporation. In a $200 million facility, 600 tons of coal are pulverized and hydrogen added under heat and pressure to turn the coal into 1,800 bbl/d of liquids that can be converted to fuel oil and gasoline.

Recently, the Reagan administration has attempted to part company with the original scheme of the Carter government. President Reagan first campaigned actively against the Synfuels Corporation and then attempted to reduce its budget. The new line is to permit the corporation to exist, but to alter its remit substantially so that government involvement is minimized, and synfuel project financing will work in such a manner as to favor major oil companies who also have access to private funding. One of the results of this change in direction was that the Japanese and West German governments decided to withdraw their financial assistance from several Synfuels Corporation projects, and private firms in these countries have indicated that they will follow suit.

There is little or no question that synfuels are an idea whose time has come. Former President Carter must be given full credit for understanding this and also for understanding that where something like energy is concerned, the marketplace cannot provide an optimal solution because of such things as international cartels and uncertainty. The government must be involved. As with so many things in the Carter administration, however, there was failure in method or mechanics, and it may have been as well that Mr. Carter was returned to private life before complete disillusionment set in with organizations like the Synfuels Corporation and the Department of Energy.

Underground Coal Gasification

Experiments with underground coal gasification date back at least a hundred years, when they were carried out by both the French and the English. Extensive work was also done in the Soviet Union during the 1930s, since it was V. I. Lenin's opinion that it would abolish the misery of underground coal miners in that country. At the present time progress in this field is slow, but in the past four years various firms and individuals in Belgium, West Germany, and the United States have insisted that it is commercially feasible to burn coal underground, and to recover and process the coal gases that are thereby created. Here it should be emphasized that the principal problem seems to be in controlling the burn, and as a result the gas product requires a great deal of expensive upgrading before it can be used. Still, it is claimed that in one 15-square-mile area in Northern Wyoming, there is enough gasifiable coal to support twenty-nine power stations of 1,000 megawatts each for twenty years (*The Economist,* January 28, 1980).

In the previous discussion of coal gasification, it was pointed out that several grades of gas were obtainable, including a gas with a very low heating value. The same is true for underground gasification. The basic gas produced will only contain 130–150 Btu of energy per cubic foot, as compared to 1,000 Btu for natural gas. Since it is uneconomical to pipe gas with this heating value an appreciable distance, it would have to be burned on site in power stations to generate electricity; but this unfortunately ties the economics of the gasification system to the economics of the electrification system, and electricity generation has fallen on hard times. On the other hand, as in the coal gasification schemes mentioned above, burning the coal in oxygen instead of air increases its energy content, although in most situations this will make the gas too expensive for generating electricity. On the other hand, it might be useful as a chemical feedstock.

The U.S. approach, based on the availability of huge resources of subbituminous coal that shrink when heated and thus allow air to seep through and stoke up the underground fire, involves drilling two types of wells. A fire is lighted beneath the shaft (or shafts) from which the gas is to be extracted, while air (or—for example—oxygen) is pumped down the other shaft. The fire seeks the source of air seeping through the coal and burns a path toward it. It is at this point that gasification begins. Despite the fact that research on underground gasification goes ahead steadily, and the basic process undoubtedly works, it seems unlikely that this technology has a real future before the next century. Not only does the technology need some fine tuning, but large coal fields must be discovered whose characteristics are favorable for gasification, and in addition there must be a direct or indirect demand for this gas. This is asking for a great deal—even in the long run.

Petrochemicals and Coal

The petrochemical industry is one of the largest in the world, with sales in the OECD of over $75 billion in 1983, excluding the sale of fuels.

In the nineteenth century, many materials for the chemicals industry were derived from coal tar, since coal is rich in the carbon, hydrogen, nitrogen, and oxygen needed to produce petrochemicals. This technology was abandoned when, after the World War II, large quantities of oil became available. Oil was just as cheap as coal and easier to transport. Unlike coal it was not a solid, did not contain a lot of ash-forming substances, and its carbon molecules are linked in simpler ways than those of coal.

The way coal is expected to return to the petrochemical scene is via gasification. Once syngas is available, it can—at least in theory—be turned into synthetic oil or methanol that can be used directly as chemical feedstocks or turned into methane, which, in turn, can be turned into methanol from which chemical feedstocks can be obtained. Methanol can be used as a chemical, feedstock, or fuel.

Natural Gas

Natural gas has been called, and justifiably, the prince of hydrocarbons. The French expression *gaz fatal* was often used at one time as a result of the large number of blowouts caused by the expansion of gas associated with oil, but this type of problem appears to be passé. The main shortcoming with gas today is that there is not enough of it; and particularly not enough of it within the territorial boundaries of the nonsocialist industrial countries.

Whenever and wherever gas has become available in the vicinity of potential consumers, the speed with which it has been adopted has been phenomenal. As explained in my book, *The Political Economy of Oil,* the discovery of large amounts of natural gas completely changed the energy picture in Holland; and great changes also took place in the United States and the Soviet Union. The Groningen gas field was discovered in the late 1950s, and by 1970 it was supplying 45 percent of all the primary energy consumed in Holland. In the United States, natural gas's share of energy consumption went from 15 percent in 1945 to 32 percent in 1965 (although it began declining soon after); and given the huge quantities of gas now becoming available by pipeline from the North Sea, the Soviet Union, Algeria, and perhaps eventually Iran, Western Europe may also be on the threshold of a new energy dawn. Here it should be emphasized that because of its cleanliness—a quality it does not share with coal and oil—natural gas may be the ideal energy medium for central and northern Europe.

Since 1973 the amount of gas burned to provide power, heat, and light, and used as a feedstock in various industrial processes, has increased by more than 40 percent on a global basis. Today natural gas supplies about 20 percent of the world's primary energy consumption. On the other hand, the amount of natural gas entering into international trade only amounts to about 12 percent of total world gas consumption, or 182 billion cubic meters (182 Gcm) in 1982. The average rate of growth of natural gas consumption, worldwide, amounted to 6.50 percent per year between 1960 and 1980, but about 16 percent per year in Western Europe during the same period. Up to 1973 the global consumption of natural gas increased by approximately 7.3 percent per year, which was almost as high as the rate of growth of crude oil consumption (which was 7.7 percent per year). Growth rates have, of course, fallen considerably since 1973, and by the beginning of 1983 may have declined to 3 to 3.5 percent per year; but it has been surmised that in the event of a return to normality in the world economy, the potential exists in Western and Eastern Europe, as well as Japan, for appreciable increases in the consumption of all categories of energy.

In general, natural gas is found in an environment similar to that of crude oil, and on occasion it has been called gaseous petroleum. The hydrocarbons of natural gas are, however, lighter and less complex than those of crude oil; and natural gas occasionally contains water and gases that are not hydrocarbons. Although many people believe that gas and oil are found in reservoirs or huge underground caverns (and this terminology is in general use), in truth they originate in water-coated pore spaces in rocks—for the most part sedimentary rock classified as organic shale. This shale, which originated as the remains of prehistoric plants and animals, was cooked into oil and gas by heat, the pressure of the earth acting over millions of years, and various chemical reactions.

Natural gas from a well consists mainly of methane (85 percent), heavier hydrocarbons collectively known as natural gas liquids (composed of ethane, propane, butane, pentane, and some heavier fractions), water, carbon dioxide, nitrogen, and some other nonhydrocarbons. Before dry natural gas can be distributed to consumers, some undesirable components must be removed and, by decreasing the share of heavier hydrocarbons, a uniform quality attained.

The last-mentioned operation takes place either at the gas well itself or in special installations. It is at this point that the natural gas liquids (NGL) can be separated out. (NGL should not be confused with liquefied natural gas (LNG), which to a considerable extent consists of methane and ethane.) It is becoming commonplace that in those regions where the availability of natural gas is greater than the absorption capacity of local markets, and for one reason or another it is not possible to transport the gas to foreign markets in its original form, to process a great deal of natural gas into gas liquids. The most important constituents of NGL are butane and propane, and in liquid form these are called LPG. In many countries LPG is sold under the name gasol or bottled gas. LPG is free of sulfur and most other contaminants and can be used in place of natural gas and as a replacement for such things as naphtha and gasoline.

Some hydrocarbon deposits contain oil but no gas, while others—where the cooking referred to previously continued until the hydrocarbons were reduced from liquid oil to molecules of gas—contain gas but no oil. Naturally, a very common arrangement is the presence of gas and oil in the same deposit, and in this situation the gas is referred to as associated gas. It is believed that gas can be found at a much greater depth than oil, and a new drilling technology is now under development that may be capable of reaching gas at depths in the vicinity of or greater than 30,000 feet, which represents the frontier of present technology. If, or most likely when, this technology permits downward probes of between 30,000 and 50,000 feet, it is expected that world reserves of gas will rise dramatically. In addition, it is thought that a great deal of unconventional natural gas can be found in coal seams, tight sands, deep basins, beneath hydrate layers that form under permafrost and the floor of the seas, and in offshore faults. About 40 to 45 percent of today's gas reserves are found in association with oil.

Since the 1860s trillions of cubic feet of associated gas have been flared (that is, burned up in the air at the site of oil wells), and it was not until the 1930s that the first major U.S. pipeline was constructed to carry gas as a byproduct of oil. Still, in 1978, about 200 billion (10^9) cubic meters of oil field gas was flared out of a total world production of 1,500 billion cubic meters (1,500 Gcm), with another 100 Gcm reinjected to raise the pressure of various oil deposits and thereby increase the amount of oil that can eventually be removed. Since 1,000 cubic meters of natural gas (1,000 cm = 1 kcm) has the energy content of 0.868 tonnes of crude oil (0.868 t), and one barrel of oil per day (1 Bbl/d) is the equivalent of 50 tonnes of oil per year (50 t/y), then 200 billion cubic meters (200 Gcm) of gas is the equivalent of 3.47 million Mbbl/d. This is calculated immediately from 200,000,000 (kcm) $\times$ 0.868/50 = 3.47. This is more than the total crude oil output of the North Sea. At least two-thirds of the gas flared belonged to the OPEC countries, but now gas-gathering schemes (such as the one being brought to completion in Saudi Arabia) will collect this gas and begin to use it as an input and as a fuel in petrochemical plants and perhaps other industrial installations.

Before continuing, a few more things must be said about units or equivalencies. Thermally, or in terms of heating values, 1,000 cubic feet of natural gas is the equivalent of 0.179 barrels of crude oil. In addition, 1,000 cubic feet is equal to 28.3 cubic meters, or 1 cubic meter = 35 cubic feet. Thus, 1,000 cubic meters of natural gas is approximately the equivalent of (1000/28.3) × 0.178 = 6.3 barrels of crude oil. This is, as always, an average value. It is also useful to know that one cubic foot of natural gas has a heating value of 1,035 Btu, which in turn is equal to 1.055 million joules, and so 1,000 cubic feet of natural gas has a heating value of approximately 1 million Btu. Price is often quoted in terms of million Btu—for example, $2.50 per MBtu. Until a few years ago this was the average price of gas in the United States, and since this is the equivalent of about $14 per barrel, and on the average oil was selling for about $30 per barrel at the same time, it is easy to understand the popularity of gas. Natural gas (supplied by the Soviet Union) was also well received in Finland for a long time and for the same reason. In 1979 Finland bought 993 million cubic meters of Soviet gas for $61 million; in 1980 imports dropped to 925 million cubic meters, but the total price was $111 million. The Finns were not very happy with this price increase, although relative to the world price of energy it was not so bad—925 Mcm for $111 million dollars gives a price of $0.12 per cm of gas. Multiplying this by 28.3 cubic meters (per thousand cubic feet) gives a price of $3.396 per thousand cubic feet or, approximately, $3.396 dollars per MBtu. Also, since 1,000 cubic feet of natural gas is the equivalent of 0.178 barrels of crude oil, on the average, the equivalent oil price is then 3.396/0.178 = $19 per barrel. A similar calculation will reveal that this comes close to doubling the former price, but it is still under thermal parity with oil.

There is at least 75 percent as much usable energy in known global reserves of natural gas as can be found in corresponding reserves of crude oil; and equally as important, reserves of natural gas are being exploited at only about half the rate of crude oil. Furthermore, percentagewise, the ratio of resources of natural gas (that is, hypothetical or potentially exploitable reserves) to exploitable reserves is larger for gas than for oil, which suggests that it may be considerably easier to locate new deposits of gas in the coming years than new deposits of oil. At the present time the principal uses of natural gas include electric power generation, hydrogen production, residential and commercial space heating and water heating, desalinization, as a feedstock for the production of petrochemicals, and reinjection for enhanced oil recovery. Natural gas reveals no particular unity insofar as global use patterns are concerned. The EEC uses about 12 to 14 percent of its gas consumption in power generation, with the figure for Italy being 10 percent and for Germany 24 percent. In both Western Europe and the United States, the premium use of gas is in the residential sector; but in Japan this sector still underutilizes gas, even though more than 40 percent of Japan's households are connected to the existing gas grid. Instead, more than 70 percent of Japan's consumption of natural gas goes to power generation.

The principal disadvantages of gas are probably associated with the huge investments required to produce, process, and transport this resource. As a result, long-term commitments—sometimes involving contracts with a running time of more than twenty years—are essential before the exploitation of even rich gas fields can commence. This observation is especially true when projects involving liquefied natural gas are under consideration; but pipelines can also be extremely expensive. The IEA has estimated that capital investment for a 48-inch diameter pipeline from North Africa or the Middle East to Central Europe, delivering 18 billion cubic meters per year (18 Gcm/year) would cost about $1.3 billion per 1,000 kilometers, including compressor stations. On the average, marine pipelines cost more than three times as much as onshore conduits, depending on water depths; but it is also true that substantial economies of scale can be realized as the volume of gas being piped is increased. A prominent disadvantage of pipelines is that they are inflexible to demand changes once pipe has been laid; and in addition pipelines outside the consumer country are, at least in theory, subject to interference by cantankerous neighbors. For example, the Trans-Mediterranean pipeline between Algeria and Italy (which cost Italy $3 billion) was not used for almost two years due to a dispute about price. Some people argued that had Italy been buying liquefied natural gas, then it might have been possible to switch suppliers; but given the investment costs of LNG, and the feasibility of obtaining natural gas by pipeline from the Soviet Union, many economists do not share this opinion. According to *Ocean Industry* (October 1979), the total capital costs for a fairly large LNG project came to $11.3 billion: $3.9 billion for liquefaction facilities, $4.5 billion for LNG tankers, $1.8 billion for terminals, and $1.1 billion for pipelines. As far as I can tell, the average cost of LNG from North Africa is about $5.5/MBtu—which is considerably higher than the price of Soviet gas.

At the present time natural gas accounts for just under 20 percent of total primary energy use in the industrialized countries. It will almost certainly become more important in Europe and the Pacific regions, where its share may decline by one-half before the end of this century. As will be explained later in this section, this is due to the simple fact that the United States is running out of conventional natural gas, although some people like to claim that it is because the price of gas has been kept artifically low, and the price mechanism has not been able to work. Unfortunately, however, the price mechanism cannot bring gas up from thousands of meters under the ground when none is present. Accordingly, drastic rises in the price of gas (and in some cases the profits of gas producers) have not been able to increase the U.S. reserves of gas. The same story, of course, is true for oil. Table 5–1 gives some information about world gas reserves and production. These figures are in cubic feet (which can be converted to cubic meters using the transformation 1 cubic meter = 35 cubic feet).

To a considerable extent, the figures in table 5–1 are self-explanatory; but it should be understood that in addition to proven reserves, it is necessary to consider *ultimate* or *hypothetical* reserves. These have been estimated at about 300 trillion cubic meters (300 Tcm = 10,000 Tcf), which means that in heating value they are roughly equivalent to 2 trillion barrels of oil—which happens to be the consensus estimate of the total quantity of recoverable oil originally found in the crust of the earth.

As indicated in table 5–1, the importance of natural gas varies widely between regions, and significant deviations from the present pattern of production (and consumption) could take place by the year 2000. Certainly, the volume of gas used in the OECD could increase by a very large increment, particularly if the huge supplies of the Soviet Union (and perhaps also the Middle East) can be exploited. Although there is considerable scope for expanding gas production in the OECD, local production will supply a diminishing proportion of OECD gas consumption in the coming years. Norway, of course, will have to be the key actor in any expansion of Western European production (as will be explained later in this section), since the estimates of total recoverable gas reserves in the British section of the North Sea are not particularly impressive. In fact, the gas fields in the southern sector of the British North Sea are already in decline. On the other hand, expectations are that, in the long run, a pipeline of several thousand kilometers can be constructed (perhaps through Sweden and Denmark) that will facilitate the entry of large quantities of Norwegian gas into the European pipeline network. The first part of table 5–2 shows some recent export patterns for pipeline natural gas.

One general comment is necessary on table 5–2 before turning to the subject of liquefied natural gas. The United States, for the most part, exports gas to Canada, while Canada in turn exports to the United States. As it happens, this has been the optimal arrangement because of the location of deposits in the two countries. There has also been considerable conjecture about Alaskan LNG being shipped to Japan.

Table 5–1
Proved Reserves (1984) and Production (1983) of Natural Gas
(billion cubic meters)

	Proved Reserves, 1984 (Billion Cubic Meters)	*Percent of World Reserves*	*Production, 1983 (Gcm)*
North America			
Canada	2613	2.9	71.34
U.S.A.	5645	6.2	450.20
Caribbean & Latin America			
Mexico	2180	2.4	31.11
Trinidad	430	0.5	3.55
Venezuela	1545	1.7	16.25
Others	117	1.7	16.25
Other Latin America			
Argentine	680	0.8	12.55
Others	479		
Middle East			
Abu Dhabi	2700	3.0	6.50
Iran	11380	12.6	8.90
Iraq	821	0.9	0.65
Kuwait	1030	1.2	4.50
Qatar	3400	3.8	4.73
Saudi Arabia	2188	2.4	5.53
Others	917		
Africa			
Algeria	3155	3.5	35.59
Libya	555	0.6	4.05
Nigeria	1370	1.5	2.00
Others	752		
Western Europe			
Holland	1927	2.1	72.98
Norway	2039	2.3	24.42
United Kingdom	712	0.8	39.53
Others	662		
Far East			
Australia	945	1.0	14.16
Thailand	240	0.3	1.61
China	800	0.9	19.80
Indonesia	1000	1.1	20.83
Malaysia	1400	1.5	3.70
Pakistan	510	0.6	9.72
Others	1495		
Eastern Europe			
Soviet Union	36000	39.9	535.95
Others	638		
	World Total	90325 Gcm = 1547 Tcf	

Source: *The OPEC Bulletin, The OPEC Quarterly Review* (various issues); IEA Documents.
Note: These are proved reserves. In the case of, for example, Norway, estimated reserves are higher.

Table 5–2
The Export and Mode of Transport of Natural Gas, 1981
(billions of cubic meters)

Exporting Country	*Pipilines*	*LNG*
USSR	57.48	
Netherlands	42.02	
Norway	26.16	
Canada	21.61	
Mexico	3.05	
Afghanistan	2.85	
Bolivia	2.18	
West Germany	1.52	
Rumania	0.30	
United States	0.12	1.45
Indonesia		11.75
Algeria		7.13
Brunei		7.05
Abu Dhabi		2.65
Libya		0.73
Total	157.29	30.76

Source: *The Petroleum Economist* (various issues); *The Oil and Gas Journal* (various issues).

Liquefied Natural Gas (LNG)

For some time now the opinion has existed among specialists in the economics of natural gas that the trade in LNG will eventually outstrip that of natural gas carried in pipelines. The basic reason for this is that the major gas-consuming countries will be unwilling to turn their backs on the huge amounts of natural gas located in regions where it cannot be economically exploited by pipeline.

Baseload LNG plants represent a relatively new and highly capital-intensive form of energy processing. A typical LNG facility requires a liquefaction plant to convert gas to a cold liquid, which is then transported to the consuming country in cryogenic tankers, and regasified at or in the vicinity of the final market. Depending on the distance between the gas source and the final market, an LNG system can be quite expensive; however, at distances greater than 8,000 kilometers—which some experts claim can be reduced to 5,000 to 6,000 kilometers if the cold in the liquid can be utilized at the point of regasification—the high cost of liquefaction (which is principally due to the relatively low efficiency of this process) is outweighed by the comparatively low cost of shipping gas over this distance. In addition, very long term contracts (fifteen to twenty years) featuring indexed prices are commonly employed in transactions involving LNG, thus reducing some of the risks associated with a transaction that is spread over a very long time span. One thing that must be emphasized, however, is that (considered in heating values) the cost of transporting gas by ship is much higher than transporting oil by ship.

The first contracts for LNG originated in 1964 between Algeria and France, and Algeria and England, and since that time trade has steadily increased. Today Algeria is probably responsible for more than 30 percent of LNG exports. Several other countries have shown an intense interest in becoming involved in this activity as soon as possible, but for a number of reasons—particularly the deepening world recession—new projects are being postponed or cancelled; however, it has been said that a once highly publicized Nigerian LNG scheme is mostly sheer nonsense from an economic point of view and would be less profitable than a 4,000 kilometer pipeline linking West Africa and Southern Europe.

Western Europe and Japan are the main customers for LNG at the present time, although the United States will undoubtedly be importing more since indigenous gas reserves are falling. In both Europe and Japan environmental considerations seem to dictate that plans to import large amounts of coal should be reassessed. In addition, another incident of the Three Mile Island variety could make the expansion of nuclear power politically impossible in many European countries and perhaps also in Japan. It should not be forgotten, however, that tankers carrying LNG can also be dangerous in the event of such things as a fire. In 1944 LNG tanks in Cleveland, Ohio, ruptured, and in the ensuing fires 133 people were killed and a large number injured. Environmental groups or persons concerned with this type of problem have been known to complain of the safety standards existing at ports where LNG is received or shipped.

A few comments can be made on some of the LNG exporting countries shown in table 5–2 before turning to the next topic. Libya's gas deposits are too far from Europe to make the export of pipeline gas feasible at the present time, but an LNG plant at Marsa el Brega began operation in 1973 and supplied substantial quantities of gas to several countries, particularly Spain and Italy. However, it has been said that operations at this installation will soon be phased out. Iran has also largely ceased to exploit its huge deposits of natural gas and seems to have completely cancelled plans to build up a huge LNG industry. I have no doubt, however, that eventually a large part of this LNG potential will be realized.

Abu Dhabi is another exporter of LNG, with all exports at present being sent to Japan. Japan is also a major purchaser of LNG from Indonesia and Brunei. The price that was paid for this LNG in 1983 was, on the average, $5.85 per MBtu, and in the Indonesian case is indexed to the prices of nineteen types of Indonesian crude oil. Other potential suppliers of LNG to Japan include Australia, Malaysia, and Thailand. It should also be mentioned that the joint Japanese-Soviet exploration effort off Sakkalin, which was mainly directed at finding oil, seems to have located considerable gas. I have been told that the Japan National Oil Company is looking for export markets for this gas, but to my mind the natural market is Japan.

Middle Eastern countries like Qatar, Saudi Arabia, and Kuwait are also potential exporters of LNG. However, it appears that these countries also have serious plans for industrial development, and so gas is also considered a potential feedstock for petrochemical installations. As a result it is not clear that investing in expensive LNG projects is an attractive proposition at the present time, although for a number of reasons it seems likely that, before the end of this century, these countries will be heavily involved in the LNG trade.

Natural Gas and Western European Energy Security

The expression "energy security" immediately starts people talking and thinking about what the Soviet Union would or would not do if it supplied a large percentage of Western Europe's natural gas, when actually the issue is much more complicated. Western Europe is an energy-poor, nonfuel minerals-poor, and to some extent overpopulated area that is on the road to serious economic and social problems if large supplies of comparatively inexpensive energy cannot be made available over the indefinite future. Energy security is actually a matter of availability, and the political objections to the USSR's sale of gas to Western Europe completely misses the point. Consequently, in the ensuing discussion, no great amount of space will be devoted to the pros and cons of expanding Soviet deliveries of gas to Western Europe. As far as I am concerned, Soviet gas is just one source of gas out of many, although the price policy of the Soviets strikes me as being much less aggressive than that of Algeria or Norway.

One of the most important characteristics of the Western European gas economy is the large amount of gas that is found within the community, particularly in Norway, Britain, and Holland. Because of this gas, and the existing and potential possibilities of the European gas transmission system (which extends from the Baltic to the Mediterranean, and from Austria to the Atlantic Ocean), Western Europe is in an extremely good position at the present time to be receptive to large imports of gas. By this I mean that if large amounts of foreign gas enter the Western European market, and were abruptly withdrawn, an almost immediate substitution could be made from local supplies. For example, the Groningen field in the Netherlands still retains the features that make it possible to develop considerable surge capacity—that is, to increase flows very rapidly in a short period of time. As a result it might be a good thing for Western Europe as a whole, though perhaps not for Holland, if this natural gas deposit were depleted at a somewhat slower rate, and thus retained an appreciable surge capacity until all the security problems associated with an increasing use of natural gas have been solved to the satisfaction of all concerned parties. It would be extremely difficult to develop an adequate surge capacity from Norwegian fields due to their location and the complexity of their pipeline networks. (On the other hand, even if Norway cannot develop an impressive surge capacity, there may be enough gas in Norwegian deposits to supply all of Western Europe's import requirements for the next twenty years). However, another important security measure is gas storage. Gas can be stored in underground caverns or, more suitably, depleted gas fields; it can also be stored in structures known as peak-shaving plants. Storing gas is expensive, but the Groningen field in Holland is ideal for this purpose. It is therefore possible that, as the Groningen field is depleted, other countries may decide to rent storage space there for their emergency supplies. Similarly, if security is a real problem, and a slower depletion of Dutch gas deposits is optimal for Western Europe than the Dutch think suitable, compensation should certainly be paid to Holland.

Taking all this into consideration, it is clear that Western Europe can afford to consume a very large amount of foreign energy materials, particularly if the price is right. Soviet natural gas, for example, is relatively so inexpensive, so plentiful, and the means for its delivery so efficient, that arguments for its rejection must be based on political considerations—or somewhat recondite economic considerations. President Reagan has tendered his reservations about this gas, since it is a policy of his government to deny the Soviets access to the sizable revenues that large sales of natural gas would engender. (The Central Intelligence Agency has estimated, mistakenly I believe, that the Soviets will earn more than $11 billion per year from the sale of natural gas by the late 1980s, and this money would undoubtedly come in handy since there is some doubt as to how long they can maintain their present exports of oil. Exports of crude oil came to 1.3 Mbbl/d in 1982, and together with exports of oil products gave the Soviet Union revenues of about $16 billion. In 1983 the Soviets delivered 19.4 Mtoe of natural gas out an estimated annual contract volume of 21.4 Mtoe. In 1984 supplies will be received under the new contracts for gas from Western Siberia, and apparently contracts have been signed for just under half the annual capacity of the new pipline—that is, 16 Mtoe out of 33 Mtoe. Accordingly, if it were desired, the Soviets could deliver another 17 Mtoe without further investments in transmission facilities.) I would like to tender my opinion at this time, however, that the principal reason for the attempted boycott of Soviet gas was not merely President Reagan's pique at the Soviet behavior in Afghanistan, or his well-documented disgust with the Communist system in general, but a quite understandable desire to force the Soviets to continue exporting crude oil in order to earn the foreign exchange they have come to need. This Soviet export of crude oil is important for maintaining the downward pressure on the oil price. There has also been talk that President Reagan was attempting to provide some help to U.S. coal exporters, in that increased sales of natural gas will inevitably reduce the revenues of these coal exporters.

As is fairly well known, President Reagan's attempt to deny pipeline technology to the USSR was not accepted by some of the most conservative governments and personalities in Europe (for example, Franz Joseph Strauss), and originally by many key executives of his own treasury, foreign, and commerce departments—whose directors were reputedly not brought to heel until they were treated to a strong dose of the chief executive's dialectical talents in the course of a working dinner at the White House. But even so the U.S. General Accounting Office, an official agency, released a study in which it emphasized that no simple alternatives to Soviet gas are available for Western Europe. Furthermore, the Congressional Office of Technology Assessment (OTA) maintains that the trade embargoes promoted by the United States to punish the Soviet Union for various alleged transgressions against international law have damaged the U.S. economy more than that of the Soviets. What the study did not say, however, is that if Western Europe had followed the recommendations of the Reagan government to the letter, the most severe punishment would have been suffered by the European economy. This would have involved running the risk of having to purchase replacement energy supplies at a price almost 20 percent higher than that offered by the Soviet Union; losing out on equipment sales worth many hundred millions of dollars (and perhaps billions of dollars), as well as the interest on loans to the Soviet Union; and creating a psychological climate that could have resulted in the future loss of Soviet markets. Something that should be stressed here is that while the economies of Western European countries could undoubtedly survive a reduction in business dealings with the Soviet Union, this arrangement could be disastrous for individual companies whose profits and prospects are already under considerable pressure and who are awaiting the coming onslaught on their markets that will almost certainly be made in the fairly near future by industries in the newly industrializing countries of East Asia and perhaps elsewhere.

Another potential supplier of large amounts of gas to Western Europe is Algeria. As it happens, Algeria is the champion of a so-called *prix-*

juste, or a price that is ostensibly optimal for producers and consumers in that it will encourage conservation by consumers, while enabling producers to replace their limited stocks of gas by other forms of capital—such as industrial establishments or educational systems. On various occasions this prix-juste has been calculated as being the same as that of crude oil, measured in terms of the heating equivalent; while on other occasions it has been declared equal to the price of synthetic gas, which means that in terms of oil it would cost at least $40 per barrel. In March 1980, the Algerian State Gas Company (SONATRACH) unilaterally doubled its LNG prices to France and the United States (that is, the El Paso Gas Company), and when the new price was declared unacceptable to these importers, trade between these market participants ceased. Similarly, Algeria unilaterally cancelled LNG export projects involving West Germany and Holland; and also it did not renew a contract with the United Kingdom because of the impossibility of agreeing on a new price. Furthermore, as already mentioned, a price dispute delayed utilization of the Trans-Mediterranean pipeline, which was built at great expense to both Italy and Algeria. Generally speaking, in the great world of natural gas importers, word was passed around that the Algerians were unreasonable.

What must be understood here, however, is that Algeria was not unreasonable in its demands. They were, on the contrary, quite reasonable—given the existing world supply-demand pattern for gas at the time Algeria was demanding an increase in gas prices. What happened to start gas flowing in the Trans-Med pipeline, and Algeria to sign new agreements concerning LNG with Gaz de France, Distrigas (USA), and Trunkline (USA) was the potential appearance on the European gas markets of huge amounts of comparatively inexpensive Soviet gas, and the possibility of bringing forward, perhaps by several years or more, exploitation of some of the northernmost Norwegian fields. In these circumstances the Algerians had no choice but to show that they, in reality, employed the same reasoning as the other vendors of natural gas or gas substitutes.

By the same token Italy, which now pays $5.40 per MBtu for Algerian gas (as compared to $4.75 per Mbtu for Soviet natural gas), hopes to be in a position to reduce the price of Algerian gas in 1985 when its contract (which has a running time of twenty-five years) comes up for review. Among other things, the Algerians will be reminded that the Soviet pipeline has arrived in Central Europe, and that the onshore Italian portion of the Trans-Med line, which was intended to carry gas from south to north, can also transport gas from north to south. In 1982, Italy's total consumption of natural gas came to around 26.7 billion cubic meters, of which 13.2 Gcm was produced domestically, 8.6 Gcm came from the Soviet Union, and 4.9 Gcm from Holland. In 1984, Algeria should supply Italy with about 7 Gcm, and plans are to take 9 Gcm from the same source in 1985. Shortly thereafter Italy should be taking its full contract amount of 12.36 Gcm via the Trans-Med pipeline; and there is even talk of eventually raising this figure to 24 Gcm/year and using the pipeline to transport Algerian gas to potential customers in Central Europe. Given these possibilities, Algeria may become more amenable to reducing the price at which it sells gas to Italy. Between 1964 and 1983 Algeria has signed contracts calling for sales of between 52.8 and 55.3 Gcm/yr of natural gas. Eleven Gcm/year are on contracts that have been suspended—1 Gcm with British Gas and 10 Gcm with El Paso (USA); while on the contract with Enagas (Spain), only 1 Gcm/year is being delivered of a contract amount of 4.5 Gcm/year. In addition to Italy, the major buyer of Algerian gas is Gas de France, which takes 9.6 Gcm/year of LNG from Algeria on contracts that were renegotiated in 1982. Distrigas and Trunkline/Panhandle of the United States together take about 5 Gcm/year of Algerian LNG. The average price at which Algerian gas is being sold is probably still over $5 per MBtu, and the estimated netback on this gas has been estimated at about $4 per MBtu by Peacock (1983). Thus, with all contracts observed, revenue from these deals would amount to between $4 and 5 billion per year.

Next, present estimates of recoverable gas reserves in Norway are 3,000 Gcm; and just now Norway is supplying Western Europe with about 14 percent of its total supplies of natural gas or about 26 or 27 Gcm/year. This gas moves through two pipeline systems: one connecting the Ekofisk field to Emden in West Germany, and the other connecting the Frigg field to St. Fergus in Britain. This estimate of reserves is definitely a minimum, however, and actual supplies are probably much higher. For example, some very promising fields in the northern part of the Norwegian sector of the North Sea have only received a superficial examination. In the following section, the numbers in parentheses, following the name of the field, are 'block' numbers assigned by the Norwegian government.

Output from the Ekofisk field is slowly decreasing, while the Frigg field appears likely to be depleted by 1995. However, the Statfjord, Heimdal, and Gullflaks fields should be coming on stream from 1986–1987 on, and they will be able to use the excess capacity in the Ekofisk pipeline caused by depletion of the Ekofisk field. The next fields due for development are the Sleipner, containing exploitable reserves of over 200 Gcm; and Oseberg (30/6), which is basically an oilfield, although it contains gas reserves of at least 55 Gcm. Sleipner is regarded as a bridge between the rapidly depleting fields in the southern sector of the Norwegian North Sea and the gas fields farther north; and the Norwegians are unwilling to enter into negotiations involving the largest of all the fields, Troll (31/2), until contracts have been settled for Sleipner. Present intentions seem to be to sell 10 to 15 Gcm/year from Sleipner at a price of $6 per MBtu, beginning about 1991; but it is difficult to see just who will be willing to pay that kind of money for natural gas, given the amount of gas that will be available about that time. Still, Shell Ltd. has declared the Troll field commercial, and that firm has been very conservative in its pronouncements about North Sea reserves.

In the discussions about Soviet gas between representatives of the Reagan government and the governments of Western European gas importers, the Troll field was frequently cited as being the principal alternative to the big Soviet gas deposits at Urengoj and Yamburg. This contention may eventually turn out to be true, but there are a number of well-known problems associated with the optimal exploitation of the Troll field. Not only does that field contain oil that could be lost if gas exploitation goes too rapidly, but water depths at some points are a fairly large multiple of water depths in other fields, such as Ekofisk and Frigg. Just as unfortunate, the gas reservoir is in a very shallow rock structure extending over approximately 700 square kilometers. As a result, a very large number of production wells will be required. Since it has been claimed that a single exploratory well in that area can cost $30 million, it seems clear that with today's technology, gas from the Troll field will be at least as costly as any being produced in the world.

The situation in Britain and Holland can also be mentioned briefly. The southern sector of the British North Sea is being rapidly depleted, and given the estimated amount of gas in the northern sector, it is just about certain that Britain will never have a substantial export capacity. The gas production of the Netherlands is irrevocably declining, and that country has introduced a conservation policy designed to stretch out its limited supplies. Exports of gas are scheduled to fall from a peak of 53.4 Gcm in 1976 to 13 Gcm in 1995. At one time it was suggested that exports should cease entirely in 1995, and Holland should save its gas for domestic use. But now it appears that the serious economic situation in Holland, and the need for export income, has led to a revaluation of this strategy, and exports are likely to continue until the next century.

By way of concluding this phase of the discussion let me say that, in my opinion, by the turn of the century the Soviet Union will have proved itself to be such a reliable supplier that only political or price considerations could lead to the rejection of the Soviets as the supplier of about one-half of Western Europe's gas. If this gas is available at a competitive price, it will be welcomed by importers. At the present time, and for the rest of the 1980s, shutoffs of Soviet or Algerian gas can be coped with for up to twelve months by the employment of emergency procedures already agreed on by IEA members; but by the end of the century, if the Soviet Union is supplying more than 30 or 35 percent of Western Europe's natural gas, this twelve months is reduced by a considerable amount. Regardless, it should be remembered that by the middle of the 1990s even some of the northernmost Norwegian fields will be ready to feed their prodigious resources into the Western European network in the event of a gas supply disruption. More important, though, is the extreme unlikelihood that the Soviet Union would be willing to permanently sever itself from the Western European energy market, which, at some time in the distant fugure, might be willing to absorb large amounts of Soviet coal and other minerals. Without going too deeply into the matter, it should be pointed out that the Soviet Union is going to require a great deal of help in one form or another if its coal industry is to be eased out of its present doldrums. But it should also be emphasized that Western Europe is also going to need a great deal of help if the present macroeconomic trend is to be broken, and among other things this means signing firm contracts for the energy required to insulate at least partially the Western European economy from the devastation that another oil price shock would cause. Although efforts throughout the industrial world to conserve energy are certainly laudable, it may be true that there is not a great deal more that can be done in this direction at the present time without laying the foundation for dangerous economic and social tensions.

Prices and Indexing

The next issue that will be treated in this section concerns the extremely important subject of indexing. If we ignore the theory of indexing, there are two practical considerations in establishing an indexation scheme. The first, and most important, is deciding which crudes, or oil products, should be involved in the indexation (or *escalation*) formula. In the deal between West Germany and the Soviet Union, the gas price is indexed to heavy fuel, heating oil, and crude oil—with the weight of the latter being only 20 percent. On the other hand, Algeria's LNG price is linked 100 percent to eight crude oils that take equal weight in the indexation formula. Thus, the recent reductions in the price of various crudes will, in theory at least, cut the price of Algerian gas by almost a dollar, which will depress it to a level well under the $6.11 per MBtu that Algeria once insisted was the absolute minimum price at which it would sell its gas. Consequently, it is often claimed that Algeria will eventually be forced to abandon its existing indexation formula for one that is more conventional. This would involve a basket of crude oils of various origins, and a medly of heavy and light fuel oils. The rate of inflation might also be given some weight.

Similarly, the crude oils in the gas price indexation clause associated with the Trans-Med project are the same as those used in the contracts for Algeria's LNG exports. In regard to the controversy mentioned previously between Italy and Algeria, it was once claimed that this indexation formula was suboptimal from the Italian point of view; but now it is the Algerians who feel misused. An example of an indexation formula reflecting 1975 price expectations is the agreement between the West Germans and Norwegians on the price of gas from the Ekofisk field. This price, p_g, in deutschmarks/billion calories, is:

$$p_g = 16.80\left[\frac{1}{2} \times \frac{A}{200} + \frac{1}{5} \times \frac{B}{275} + \frac{3}{10} \times \frac{C}{170}\right]$$

where A is the price of heavy fuel oil to customers in northern Germany, plus 60 percent of the difference between the price of a standard fuel oil and a low sulfur fuel oil; B is the price of gas oil; and C is the price (f.o.b.) of heavy fuel oil in Rotterdam. As things have turned out, there has been a very noticeable shift in demand from heavy to light petroleum products over the past few years, and this has meant a noticeable decrease in the price of gas when it is determined using this type of formula. There is no need to point out, I hope, that in situations where either the buyer or seller of gas comes to the conclusion, after the fact, that they are being treated unfairly by an existing indexation formula, suggestions are immediately forwarded for a review of the formula. Given the present world supply-demand situation, however, it seems unlikely that buyers of natural gas would be sympathetic to a revision of existing arrangements, and such is likely to be the case for some years in the future.

Another important matter that, unfortunately, cannot be treated in great detail here, involves whether gas prices should be indexed to the price of crude oil, for example, on a c.i.f. or f.o.b. basis. This is important because transportation and associated costs can be as much as eight times higher for gas as for oil, measured on a heating value basis. Thus, importers are unlikely to favor f.o.b. pricing since a slight rise in the price of oil can have drastic implications for the final price of gas.

U.S. Natural Gas

As already pointed out, the United States is the largest consumer of natural gas in the world; however, a palpable downward trend exists for gas consumption in that country, and apparently nothing can or will be done to reverse this movement. Indeed, by bringing gas prices into parity with competing fuels at a time when the energy intensity of U.S. industry is declining, the authorities have virtually ensured that the long-run demand for gas will remain moribund. It is also useful to know that, at present, 15 percent of U.S. supplies are unsold and languishing in shut-in production wells.

At the same time, the rise in gas prices encourages exploration and production. In the case of new gas developed from deep or nontraditional sources, all price controls have been removed; and for new gas discovered after 1976, prices will be completely decontroled by 1985. Old gas (discovered prior to 1976) will continue to be under federal control, but there will be periodic price increases. The policy document that spells out these rules in detail is the Natural Gas Policy Act (NGPA), which also has established twenty-two different categories of natural gas, depending on age, location, depth, and cost. Price ceilings have been established for all these categories, and federal price controls have been extended to interstate gas, thus completing a nationwide system of wellhead price regulation. So-called new gas, which has come into production since 1977, accounts for slightly more than 40 percent of the current U.S. output, and a small amount of gas from categories with high production costs has sold at wellhead prices of from $3 to $7 per MBtu, when the price of all gas from new sources averaged $2.40 per MBtu.

There are about 225,000 producing gas wells in the United States. Of these, 16,000 were drilled in 1982, as compared to 9,800 five years earlier; but this increased drilling activity has been insufficient to keep gas reserves from falling. The problem here is the productivity of the average gas well. In 1966, the average gas well produced 115 Mcf/year, while in 1983 this output had falled to 65 Mcf/year. If this trend continues, several hundred thousand new wells will be needed by 1990 if today's output is to be maintained. Needless to say, it is extremely unlikely that these wells will be forthcoming, which means that it will be both physically and economically impossible to maintain the present production of natural gas in the United States.

U.S. gas reserves are shown in table 5–3, and it has been estimated that these are adequate for at least the next decade if there are no unforeseen increases in consumption. It can also be noted that the United States receives large imports of pipeline gas from Canada, and Mexico should eventually be able to supply the United States with a great deal of gas. Given the excess supply of gas now existing in the United States, this is not a pleasant prospect for producers in that country, and to a considerable extent they are relying on the government to block exports from Canada and Mexico. At the same time, producers in Western

Table 5–3
U.S. Reserves and Resources of Natural Gas, January 1983

	Reserves (Tcf)	*Resources*[1] *(Tcf)*
Onshore		
Lower 48 states	151	179
Alaska	6	28
Total	157	207
Offshore		
Lower 48 states	33	53
Alaska	2	69
Total	35	122
Total U.S. Reserves	192 Tcf	329

Source: *Petroleum Economist*, various issues; IEA documents.
[1]Resources are defined as speculative reserves.

Canada are urging their government to permit them to lower prices so that they can export more to the United States.

The principal purpose of this subsection is to examine the pattern of demand in the United States. There has been a great deal of oil, gas, and coal in the United States since World War II, prices have been low, and so consumers have had a long time to examine the relative merits of various energy media and select those which they feel are optimal for different uses (residential, commercial, electric power, and other industrial). It must be acknowledged, though, that since 1973 prices have been out of line in the sense that they have not been at or in the vicinity of their market-clearing levels, and thus it is not strictly correct to speak of optimality. Also, since 1973, there has been an attempt by various administrations to interfere directly in the use of energy supplies. For instance, the U.S. government has passed laws intended to restrict the use of gas by industry (especially as a boiler fuel), in the interests of making more gas available for so-called premium uses—house and shop lighting and heating, or any use where a clean flame is required.

But even so, in absolute terms, the residential and commercial use of gas has declined since 1973, although in percentage terms this sector consumes about 45 percent of all the gas used in the United States. In the same vein, gas is maintaining its relative position in the industrial sector, where (in 1982) 38 percent of the total consumption of natural gas took place. In this use gas mostly competes with light and heavy fuel oil, although the upgrading of refineries to transform residual oil to lighter products will probably mean that less heavy fuel oil is available. Rather than pay higher prices for that which is available, some present buyers of this product will switch to coal or gas.

Electric utilities use about 17 percent of the natural gas consumed in the United States. Analogous to the situation in other countries, there is a slight tendency for this fraction to rise in the United States, although there is a general belief that any increase in consumption would only be temporary. As explained in chapter 7, gas is a premium fuel for so-called peak-load applications; but it could happen that gas is also carrying more of the intermediate load in the present business cycle upturn and will continue to do so unless it appears that there is a marked upswing in the growth in electricity consumption, and therefore more conventional intermediate load equipment is warranted. On this point it should be noted that gas turbines possess a flexibility that eludes most other equipment for generating electricity. They have the lowest installed capital cost, the fastest start-up time (5 to 10 minutes), the ability to burn a wide range of fuels (from natural gas to crude or residual oil), and the lowest maintenance costs.

The last topic to be taken up here deals with pipelines in the United States and the pipeline companies that operate them. Generally these companies buy enough gas from gas producers to meet what they consider to be expected demand, and their customers pay approximately the same price as they do for this gas. The profit of pipeline companies is obtained from the transportation charges on the gas. The profitability of these pipeline firms is now threatened by the fall in gas consumption (with the use rate in many interstate pipelines already below 70 percent) and by the possibility of widespread deregulation that will erode their monopoly position in the gas transportation market, and consequently lower a level of return on invested capital that has turned out to be well above the average. Some pipeline companies are meeting this possible jeopardy by moving more deeply into the production of gas—just as more producing companies are trying to extend into selling gas directly to industrial consumers (and therefore depriving pipeline companies of long-term commitments if they can). In general the trend is toward all market participants displaying a greater willingness to experiment with shorter term contracts, including more spot market sales. To my way of thinking, however, this is a passing fad, and eventually there will be a reversion to the conventional way of doing business. Another problem now being dealt with more expediently consists of the liabilities that resulted from the take-or-pay contracts that many pipelines have signed with producers (and which seemed so beneficial a decade ago when there was talk of a gas shortage). These contracts require pipelines to take as much as 80 percent of contracted volumes of gas over an extended period of time or pay for the gas even if it cannot be disposed of in the market. Contracts of this nature now keep pipelines from turning to cheaper, price-controlled gas. An increasing number of pipeline companies are now handling this dilemma by ignoring their obligation to pay for the gas, and go instead into litigation. Invariably, the matter is settled by gas producers taking 10 to 20 cents on the dollar for the gas that the pipelines do not take.

Interestingly enough, the present turbulence seems to be leading to the formation of a major spot market in gas. Brokers are organizing producers with cheaper supplies to sell, while at the same time finding customers for spot gas. One of the problems here will be to find pipeline companies with spare capacity who are willing to transport spot market gas (and there may not be a great many of these because these companies do not want to be put in the position of being played off against each other), and also to eventually establish large regional or nationwide gas pipeline grids, because without these kinds of facilities, no spot market in gas can function. By way of contrast, this does not mean that with these facilities it will be possible to build up the kind of spot market that exists for some other commodities, including oil. The problem here is that pipelines may turn out to be a much less flexible means of transport than, for example, oil tankers.

If the spot market in gas does get on its feet in the United States and captures the approximately 20 percent of the natural gas market that some people feel it can capture—which is not at all certain—U.S. consumers of natural gas may find themselves in position to reap substantial gains. But at the same times there will be many players of the gas game in the ranks of producers and pipeline companies whose worries have just begun.

Some Further Notes on North Sea Gas

After several years of decline, gas consumption in Western Europe increased to 205 Gcm in 1982, which was about 4 percent higher than in 1981. The question now is what will the future bring?

A number of rough estimates and projections are available. One of the most conservative with which I am acquainted is offered by Davies (1984), who selects two cases—a so-called likely case, with an average annual growth rate of demand of 1.6 percent for the 1983–2000 period; and a high growth case of 2.1 percent per year for the same period. In the first case the share of gas in primary energy use remains at 15 percent, while in the second it rises to 15.7 percent. Personally, I find these forecasts too low, although if Western Europe does not emerge from its present economic doldrums, then they could be quite realistic.

In the present subsection I want to pay closer attention to the North Sea than was done earlier in this chapter. It has already been pointed out that there is a queue of new fields expected to come on stream in the Norwegian portion of the North Sea, but what was not sufficiently emphasized was the steady fall in output from existing fields after 1985. The profile (for existing fields) offered by Davies results in production being almost flat from 1985 to 1990, at about 60 Gcm/year, after which there is a steady decline at about 4.55 Gcm/year until 2000, when production reaches 22.7 Gcm/year; but since additional fields will undoubtedly be brought in—mostly in the far north—there will probably be a plateau production of 102.5 Gcm by the year 2000.

Over the past two years, I have been given very good reason to believe that the Norwegian government and people are very unhappy about the new gas supply situation created by rising exports of Soviet gas to Western Europe. Norway now has the second highest standard of living in Europe (after Switzerland) when calculated in income per capita, which to this economist means that when welfare benefits are included it has the highest standard not only in Europe, but in the world. Unfortunately, though, this does not appear to be enough. What is desired by some, although probably not a majority of the citizenry, is not just comfort but ostentation; whereas the authorities—who must be concerned with the growing welfare expenditures and the future of the country after its oil and gas are exhausted—want more income to subsidize and modernize Norway's traditional industries and perhaps to establish new industries. Prior to the entry of new Soviet gas into the Western European energy picture, Norway was as hard a hard-liner on the matter of gas prices as Algeria. Since then, it has reluctantly adjusted to reality, which among other things means tacitly accepting a fall in projected gas revenues of between 15 and 20 percent due to the softening of the gas price.

While being highly concerned with the future of the continental gas market, Norway has also been busy negotiating a major deal with the British Gas Corporation. This was for gas worth £20 billion, over a thirty-year period, with most of this gas taken from the Sleipner field. A simple calculation will show that this amounts to £666.666 million per year, or about 6.7 billion Norwegian crowns at the present rate of exchange. This transaction, by itself, would furnish almost enough money to finance the Norwegian educational system over the period of the deal; but as things have turned out, the papers finalizing the deal were never signed because the British government—as distinct from British Gas—did not believe that there would be a serious gas supply gap in the late 1980s or 1990s. There is no point in my speculating on how this matter will be resolved, because apparently this is a political as well as an economic issue. The Norwegians, of course, want premium prices for this gas, claiming that it will be extremely expensive to produce (for reasons given in the previous subsection), and since the United States wants to see Norwegian gas developed as an alternative to the introduction of even more Soviet gas into Western Europe, it argues that Britain should pay this price.

One possibility that should be considered, however, is that the Norwegian gas contract is renegotiated, and British Gas takes only some fraction of the amount of gas proposed on the original contract. The remainder would be obtained from the Netherlands, whose budgetary difficulties in the light of an ivory tower attitude toward welfare expenditures and so-called development aid have now caused that country to reverse its previous gas depletion policy and seek new export contracts. The problem here is that the latter deal would connect Britain to the continental gas grid for the first time, which might eventually mean Soviet gas coming to Britain. But by the same token it might mean that free market–oriented British governments would be encouraged to permit the export of British gas to the Continent, where gas prices are considerably higher. The policy of British Gas has been to hold down domestic gas prices, and the rationale of the proposed deal with Norway is to ensure supplies of gas at prices that, when compared to present British gas prices, might appear high, but which British Gas feels will be reasonable in the light of the probable gas price level in the 1990s. The price that British Gas pays for North Sea gas from British fields has recently risen from $3.40 per MBtu to $3.55 per MBtu; and according to estimates in the British press, the price at which this gas from the Sleipner field was to be sold was about $4.45 per MBtu. This represents quite a comedown from the previous price asked for Sleipner gas (which was said to be a nonnegotiable $6.00 per MBtu), but it is understandable that the Norwegians would be willing to compromise, because in a decade or so they are going to have some problems finding buyers for gas from their Troll field, and British Gas is a highly esteemed potential customer. Let me make it clear, however, that I entertain serious doubts as to whether gas from the Troll field can be obtained at less than premium prices.

Paradoxically, while the U.S. government may be skeptical about a United Kingdom link-up with the continental gas grid, the IEA and the EEC are positive, seeing a Channel pipeline as an important factor in moderating a supply crisis, since it would allow British natural gas to be routed to the Continent.

The hydrocarbon sector now accounts for one-third of Norway's exports and 17 percent of its gross national product. It is believed that it will be contributing 20 to 25 percent to gross national product by the end of the century. It has been suggested that this forecast will not come true if it does not become more profitable for oil companies to operate in that area, but the Norwegian government is inclined to disagree. Similar complaints have been heard about the tax policy of the British government; however, many of the same people who complain about the tax policy say that the real problem is the price problem. At least one energy company has claimed that if British North Sea gas were sold at competitive prices, enough of it would be found to make foreign purchases unnecessary. Statements of this nature have been intepreted by one well-known British oil economist to mean that many energy firms are hiding their reserves until the price of gas is raised to at least match the $4 per MBtu that, on the average, prevails on the Continent.

Increased flexibility on the part of the Norwegian government should not be ruled out. The Troll field will be very tricky to exploit, since water depths are more than 1,000 feet, the sea bed is soft, and the reservoir shallow. This depth, incidentally, is greater than that at which any installation is operating today, and it has been said that to operate at this depth in Arctic waters may require an entire new technology. For instance, the present record depth for the North Sea is about 350 feet for British Petroleum's Magnus Field, and this is considerably south of the proposed Troll development. Still, work will almost certainly go ahead. According to the latest estimates, Troll gas reserves may amount to 2 Tcm, which is more than the extremely rich Groningen Field. Shell has set the target date for the first deliveries of gas from this field at 1995, assuming that somebody is willing to buy the gas at an as yet unannounced price. If not, production will not begin until much later.

Finally, a few words about Denmark, which has graduated to a North Sea energy power from its previous status as a candidate for the title of "sick man of Europe." Almost 20 percent of Danish oil production is from Danish North Sea fields, and once gas starts flowing at the production target of 1.7 Gcm/year, between 20 and 25 percent of Danish energy requirements will be provided for by domestic oil and gas fields. It has been suggested by so-called reliable sources that this figure could reach 30 or 40 percent in the 1990s, although some doubt has been expressed as to whether sufficient oil and gas actually exists in the Danish sector of the North Sea to justify this contention. A great deal of coal is being bought by Denmark at the present time to replace oil, particularly for electricity generation; and nuclear energy is not used.

Danish industrial users pay between 1.50 and 1.80 Danish crowns for a cubic meter of gas; households pay 3.3 Danish crowns. By international standards, this is very inexpensive—and much cheaper than oil. It has been rumored that the financial condition of the Danish state oil and gas company (DONG) is not as good as its directors claim, although some clever bookkeeping has enabled the auditors to certify that the company is still in the black. Fears have been expressed in Denmark that DONG will become as independent and powerful as the Norwegian state oil company (STATOIL), but given the comparatively small amount of resources controlled by DONG, this seems extremely unlikely.

Conclusions

This very long section was intended to introduce the reader to the economics of natural gas, and in particular to illuminte the natural gas economy of Western Europe. But remembering that one of the most important future tasks of mainstream economics is not just to present new ideas and information, but to get rid of the innumerable bad ideas that are all around us before they cause even more confusion, I feel obliged to conclude this analysis with the following observations.

Wilfred Prewo of the Institute für Weltwirtschaft at the University of Kiel has come to the conclusion (in "The Pipeline: White Elephant or Trojan Horse?" the *Wall Street Journal*, September 28, 1982) that there are five sources of natural gas for Western Europe that are as inexpensive as Soviet gas, among which he includes the South Atlantic and Canadian Arctic. In point of fact, however, this is the exact opposite of the way things were and are. For example, in 1982 a Canadian consortium discussing possible deals with Gaz de France, Ruhrgas, and other large European importers involving contracts of up to twenty years, was unable to provide any guarantees on price and delivery. Let me also take this occasion to express my complete lack of sympathy with those people who view acceptance of the Soviet pipeline by Western European governments as an attempt to resuscitate a half-dead détente (including the journalist William Safire in the London *Times*, August 16, 1982). As I pointed out in my lecture at Cambridge University in June 1982, few things have helped to revive détente more than President Reagan's attempted embargo with its cold war overtones. Without the embargo, which happened to be pointless in both conception and practice, the entire pipeline issue would have been a strictly business arrangement of the type indulged in by the United States with the Soviet Union while Soviet-manufactured rockets, some of which may have been fired by Soviet technicians, were being used against U.S. aircraft in the skies over North Vietnam. Furthermore, as former President Nixon has repeatedly pointed out, the more business that the USSR does with the West, the more susceptible it will be to friendly persuasion in those political matters where legitimate U.S. interests are at stake.

The question of a GASPEC is also relevant here, since I had this matter put before me in extremely aggressive tones at the 1982 Cambridge meeting of the International Association of Energy Economists by an otherwise placid and *only slightly* perplexed member of the academic–government complex. Since the Soviet Union, which is the largest producer of oil in the world, and has exported as much as 1.5 Mbbl/d to the West, did not apply for even an associate membership in OPEC, I see no logical reason for it to be interested in taking part in a gas counterpart to that organization. On the contrary, the huge export capacity of the Soviet gas industry has probably been instrumental in stopping even the consideration of forming an Organization of Gas Exporters. However, if a GASPEC is inevitable, perhaps the best way to moderate its impact is to have a huge stock of gas remaining in Western European deposits that can be turned into a flow in a relatively short time. As it happens, one of the major reasons for the success of OPEC was the restrictions on U.S. oil imports that were imposed by the Eisenhower administration in the 1950s. Among other things, the fall in imports led to a substantial reduction in domestic U.S. reserves.

The kind of bad logic and indifference that led to the devastating energy price rises of the 1970s was regrettable, to say the least; but they were not nearly so regrettable as the intellectual collapse that would be implied by a decision to send Western Europe to the waters of the South Atlantic in search of its energy salvation.

6
Coal and the Environment

This chapter is a partial survey of coal and the environment. Its purpose is to bring together, interpret, and comment on some of the latest facts and figures relevant to this crucial topic. Its justification is that, to my knowledge, a comparable overview does not exist elsewhere. However, if I am wrong in this belief, I still think that the reader will find the materials presented in this chapter useful, particularly since they are submitted in a framework of plain and unambiguous talk that is not always representative of a large part of the academic economic scene. I would also like to make it clear that this chapter can be read independently from the remainder of the book.

The conclusions to which I come can, in one sense or another, be regarded as an extention and elaboration of my earlier work on such topics as environmental problems and population; however, on this occasion I am particularly concerned with providing some insight into the economic aspects of acid rain and the greenhouse effect, taxes and emission standards as policy devices for controlling environmental deterioration, and various intertemporal themes associated with the concept of option value, among other things. For a more detailed examination of these matters, see Gollop and Roberts (1981, 1983), and Edmonds and Reilly (1983a, 1983b, 1983c). Let me emphasize, however, that I am not agitating against a return to the age of coal; I am concerned that if this is due to happen, environmental considerations are given their proper weight. By way of contrast, I am agitating for such things as the Waxman-Sikorski legislation in the United States, calling for a direct and immediate attack on environmental deterioration of the kind successfully employed by the Japanese government. As far as this economist is concerned, the message from the German forests is unambiguous.

After some thought I have decided to use a very small amount of algebraic exposition in the main part of the discussion, rather than in the appendix. These materials are kept as elementary as possible, and are included only for completeness. In all cases, however, they can be bypassed by readers who are not interested in this type of presentation.

And finally, I would like to use this opportunity to reaffirm a position stressed in *The Political Economy of Oil*. At this stage in human history, the great danger to mankind is not pollution but population. It is the exploding world population that is slowly but surely opening the door to environmental and economic disaster.

The Basic Problem: Uncertainty

Someone has said that economists tend to advise their governmental clients to maximize expected social benefits, whereas environmentalists emphasize that since expectations are not always fulfilled, optimal economic behavior invariably turns on keeping future options open. Needless to say, this is not quite the truth, because the existence of uncertainty in almost all human endeavors, together with the presence of risk aversion and the irreversibility of many economic processes, generally makes flexibility an attractive proposition for even the most dogmatic members of the economics profession.

Flexibility where fossil fuels are concerned probably means a lower than anticipated growth rate of coal consumption, at least in the near future. This issue has been confounded by a magnificent outpouring of bunkum in the form of so-called scientific analyses, but the bare bones of the matter come down to the following. The environmental effects of a large portion of atmospheric pollution will only be felt in the distant future. As a result, it often happens that uncertainty about the damage resulting from pollution cannot be resolved for years, perhaps decades, and thus policymakers operating in the present cannot reckon with this uncertainty being dispelled before serious and, in some cases, irreparable harm has been done. Accordingly, current resource use (that is, fossil fuel consumption) should be more conservative than in situations where decisions made today do not entail a foreclosing of tomorrow's alternatives. This is what the supposedly complex subject of option value is all about, and obviously it has a lot to recommend it—even within the domain of common sense.

Another item of importance for the present discussion has to do with the value of information, or the optimal amount of real resources that a rational agent would surrender in order to dispel a given amount of uncertainty. Without going too deeply into this subject, it seems evident that most of the time enough information cannot be obtained to convert uncertainty into perfect certainty at any price; but in many situations where the decisionmaker is not operating at an extreme (in the sense of being absolutely convinced of the outcome of an uncertain event), it is theoretically possible to impute a money or resource value to knowledge that permits the revision of subjective beliefs about the likelihood of future events. Furthermore, information is more valuable for someone who is very risk-averse than someone only slightly risk-averse—or risk-neutral; and, by extension, the larger the penalties resulting from making the wrong dccision, the more valuable the information. (In simple terms, if a person or group of persons were risk-averse, they would only bet a sum on money on the flip of an unbiased coin if they received more than that sum in the event they won. In other words, they are averse to accepting a fair gamble. On the other hand, a risk-neutral person would accept certain fair bets.) In the context of our present discussion, the cost of information can be measured in terms of the goods and services that have to be foregone during the period when the environmental damage that could be caused by the burning of fossil fuels was being assessed, and the use of such things as coal was being kept below the level which would have prevailed if these suspected or alleged spillovers (that is, external effects) did not exist. A few years ago this cost—whatever it was—was regarded as excessive by many businessmen, politicians, international civil servants, and other important personalities who are charged with keeping the wheels of progress spinning. This attitude seems to be in the process of changing.

As an aid to thinking about these issues, consider the present situation in West Germany, where it appears that about a third of the forests have been damaged by chemical agents derived from fossil fuels, although only a year ago the injury to forests was estimated to have been considerably less than a half of the current level. What has happened is that, as shown in figure 6–1, some kind of threshold has been reached. Thus, in the vicinity of a pollution level F^*, the damage $D(F)$ from pollutants like sulfur dioxide (SO_2) rises very rapidly. In addition, such things as SO_2 and various atmospheric particulates, which if taken individually might not be dangerous except at high concentrations, could be hazardous to human health at fairly low concentrations when combined. In the nonlinear damage function shown in figure 6–1, P represents different, particulate concentrations with $P_1 > P_2 > P_3$; and if—for example—$P = P_1$, or greater, severe damage can result if $F > F^*$.

Another thing to be noticed in figure 6–1 is that to the right of F^{**}, the flow of damage is large, although the marginal damage (dD/dF) is comparatively small. Thus the kind of economic reasoning that emphasizes marginal phenomena might draw the conclusion that economic activity causing pollution levels higher than F^{**} was relatively harmless, when in fact a socially optimal policy might call for a lower flow of

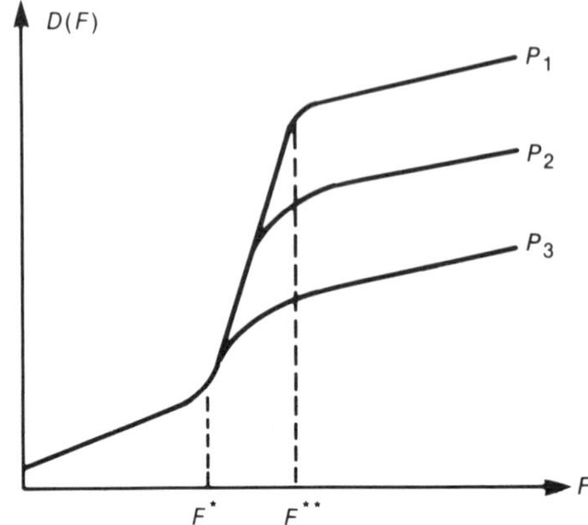

Figure 6–1. Nonlinear damage function $D = D(F)$, where D is environmental damage, F is the level of a certain pollutant (SO_2), and P is the level of another pollutant (such as atmospheric particulates)

damage, such as in the vicinity of F^*, where the marginal damage was higher. As far as I can tell, nonlinearities of this type vitiate a few of the most cherished results of neoclassical economic analysis, including some of the conventional wisdom claiming that taxes are superior to regulations as a device for curbing pollution.

Since coal-burning power stations are responsible for approximately two-thirds of the atmospheric deterioration of Central Europe, it has become clear to the German government that previous legislation designed to reduce emissions from these installations was inadequate. Furthermore, even though the price that must be paid for finding out about and reducing environmental damage is rising all the time, which increases the temptation to shift the financial burdens associated with pollution abatement farther into the future, a point has been reached where environmental decay is accelerating at such a rate as to make this kind of transfer both politically and economically impractical. As a result, on July 1, 1983, a law was passed in Germany specifying that new electrical power stations supplying over 300 MW must be capable of meeting especially rigid emission standards and that older power stations must conform to these same standards within six years. The SO_2 content of the atmosphere must fall drastically, and a program has been outlined calling for a 30 percent reduction before 1993. People like the German ecologist Herman Count Hatzfeldt want Japanese air standards adopted, which he estimates would cost about 3 billion deutschmarks per year for at least ten years.

Unfortunately for all concerned, this will probably not be enough. Emission standards must be even stricter, at least in the short run; and in fact a thorough environmental cleanup should encompass polluters of all categories and may necessitate sustained abatement programs that initially cost, for a country like the United States, at least $5 billion per year, and perhaps as much as $10 billion. Most important, where expenditures of this magnitude are concerned, microeconomic exercises proving that taxes levied directly on polluters are more efficient than emission standards (that is, regulations) completely miss the point. The opinion here is that an optimal policy must be framed in terms of the macroeconomy, and will feature standards plus subsidies to industrial polluters. Concomitantly, since the environment is a community concern, taxes to pay the subsidies should come from the entire community, not just the productive sector—where very heavy costs may eventually be transformed into lower investment and additional unemployment.

It has been suggested that European countries must reduce their output of SO_2 by 75 percent in the near future if the war against environmental deterioration is not to end in a defeat that will have catastrophic effects for coming generations and perhaps for many members of this generation. But bringing this reduction about via taxes on polluters that result in a lowering of industrial production might result in another type of debacle, since in my opinion, in the not too distant future, many industries in Europe will be struggling to stay afloat in the face of competition from the newly industrializing countries of East Asia and elsewhere.

Although it is not my intention to paint a picture of doom, I think it needs to be said that at present there is no suitable barrier between ourselves and any number of potential environmental disasters. The principal problem is that—as in the example from Germany—not only are people unaware of what is happening, but subtle forces are at work to keep them from finding out. A study by the U.S. Environmental Protection Agency (EPA) predicting a two-degree average rise in the global temperature was rejected by President Reagan's science adviser, George Keyworth, as "unnecessarily alarmist"; while a report by the National Research Council, stating that a doubling of atmospheric carbon dioxide by late in the next century would cause a temperature increase of between 1.5 and 4.5 degrees, was praised—apparently because it concluded by announcing that "our stance is conservative: we believe there is reason for caution, not panic." It appears that Dr. Keyworth was annoyed by the EPA's contention that early planning was necessary to deal with the serious disruptive effects that climatic changes almost certainly would evoke, although apparently this so-called planning involved no more than allocating money for research on how to adapt to environmental changes. It is also true that neither document made explicit mention of drastically curtailing an increase in the consumption of certain fossil fuels if the evidence indicates that this is where the problem lies, although it occurred to some observers that a recommendation to this effect might have been warranted.

What it all comes down to, I am afraid, is that many highly educated and socially conscious representatives of the scientific establishment have never heard of option value, nor for that matter indicated that they have the capacity to understand its wider connotations. For instance, if there is a sizable probability of a 4.5-degree average increase in global temperatures in the next century, it is already time to replace caution by action.

The Supply of Coal

In the light of the concluding remarks in the previous section, I see no harm in mentioning that, after the first oil price shock, I was one of the first economists in the world to advocate a large-scale return to coal as an alternative to paying for ever more costly oil. My assumptions at that time were (1) coal was, and would remain, much less expensive than oil; (2) the economies of scale in coal use were more substantial than generally realized, and thus the price of electricity generated from coal could be extremely attractive; and (3) huge quantities of coal were located in the right places (that is, the United States, Canada, and Australia). Given these advantages, it would be a simple matter to afford the high-priced investments that might be required to make coal a useful and relatively clean servant of the industrial world.

Over the past half-year or so the price of coal has fallen steadily relative to the price of most other primary energy resources, although in terms of heating value this price has been lower than these materials for almost the past half decade. But at the same time the cost of pollution-abating equipment has been rising; and we have found out a great deal about the inadequacy of a large part of this equipment, regardless of how much has to be paid for it. There has also occurred, of late, a considerable increase in the availability of natural gas that may be capable of keeping gas prices depressed in many world markets throughout most of this decade and perhaps some of the next. In fact, it is not unthinkable that gas might eventually replace oil in Western Europe as the dominant energy medium—particularly if certain political problems involving the Soviet Union are solved and the gas of the *Gulf* can be optimally exploited. Against this background, coal has turned out to be a much less appealing proposition than it seemed in the recent past. Yes, huge increases in coal production and use will definitely have to take place as a precaution against immoderate rises in both oil and gas prices; but for reasons that are made clear in the remaining sections of this chapter, the goal at the present time must be to ensure that these increases take place in an environmentally suitable framework.

At this point it should be made explicit that the story of coal is virtually the same for every industrial country, over most of the postwar world, regardless of whether we are talking about the total input of energy in the form of coal, or the industrial demand for coal, or the use of coal in the generation of electricity by so-called public utilities. From being the most prominent element of the energy picture in 1950, when it supplied more than 40 percent of total energy consumption on a worldwide basis, coal declined to less than 10 percent by 1973. The first oil price shock slowed this withdrawal, and the second reversed it by a small amount. Table 6–1 provides a rough glance into this situation; but at the same time, although coal has been eclipsed by other energy sources on a relative basis, its production has still grown by about 2 percent per year between 1950 and 1980.

Next we can scrutinize the popular fantasy that world coal resources are virtually unlimited. To get an insight into this misconception, it is useful to review the state of affairs in the United States, where approximately 30 percent of the world's coal reserves (that is, coal resources that are exploitable at present levels of economic knowledge) are located. These amount to about 113×10^9 tonnes of hard coal, and 64.4×10^9 tonnes of brown coal—which together can be called 145×10^9 tonnes of

Table 6–1
GNP, Energy Growth, Oil Growth, and Coal Growth, 1968–1980

	GNP		*Energy Growth*		*Oil Growth*		*Coal Growth*	
	World	*OECD*	*World*	*OECD*	*World*	*OECD*	*World*	*OECD*
1968–1973	5.2	5.0	5.2	5.0	7.7	7.2	1.4	−0.3
1974–1978	3.3	2.5	2.6	0.9	2.0	0.4	2.3	0.4
1979–1980	2.8	2.2	1.2	−0.5	−1.3	−4.0	4.0	6.2

hard coal equivalent. On the other hand, estimated total geological resources (that is, exploitable resources plus hypothetical resources) are $1{,}200 \times 10^9$ tonnes of hard coal equivalent, and 1380 tonnes of brown coal $(1{,}200 + 0.5 \times 1{,}380) \times 10^9$ tonnes of hard coal equivalent. Now it is universally agreed that if coal is to be the bridge between oil and renewable energy sources, there must be an enormous expansion of coal production in the United States. It is often suggested that a production increase of 10 percent per year is reasonable, which means that production would rise from somewhat less than 700 million tonnes per year at the present time to about 3,215 million tonnes per year by the turn of the century. As for the length of time needed to exhaust existing reserves, we can use the formula derived in appendix 2A; but by the same token we know that the sum of a geometrical progression can be written:

$$s = a\,\frac{(r^n - 1)}{r - 1} \quad \text{when} \quad r > 1$$

In this expression, a is the starting amount, while r is unity plus the growth rate of output. If the rate of growth is 10 percent, then $r = 1.1$. It is a simple matter to solve this expression for the length of life of reserves n, given total reserves $\bar{s}$:

$$n = \frac{\ln\left[\dfrac{\bar{s}\,(r - 1)}{a} + 1\right]}{\ln r} = \frac{\ln\left[\dfrac{145{,}000\,(1.1 - 1)}{700} + 1\right]}{\ln 1.1} = 32.2 \text{ years}$$

In fact, even if the total geological resources were assumed to be fully exploitable in the near future, we would only get:

$$n = \frac{\ln\left[\dfrac{1890000\,(1.1 - 1)}{700} + 1\right]}{\ln 1.1} \approx 59 \text{ years}$$

Accordingly, a coal age based on a 10 percent growth rate of coal production in the United States is going to be a fairly brief episode. Clearly, a 5 percent growth rate, or even less, is more attractive in many respects, since it would mean that existing coal resources would last about 100 years, which should provide enough time to fully develop fusion—if it is developable—and enable solar energy to be turned into an economic proposition.

Before beginning our examination of some of the central issues associated with fossil fuel use and atmospheric pollution, there is a peripheral item that deserves mentioning. Half the world's population employs wood as its main source of energy. Unfortunately, though, wood possesses only a modest amount of energy in regard to its volume; and in addition the combustion of wood generates a very large volume of carbon dioxide. For this reason, and to preserve the large forests remaining in the less developed countries—which along with the oceans are the most important sinks for carbon dioxide—it has been suggested by organizations like UNESCO that more coal should be used. In their opinion, coal is the energy resource that is most evenly spread out over the globe. There is no need to comment here on this kind of advice, although I would like to say that if coal use were expanded at the rate that organizations like UNESCO believe to be feasible, the march back into the age of coal will be terminated virtually before it begins (just as the oil epoch may soon be over for several OPEC countries, although they do not realize it yet), and this could be true even if the USSR is eventually able to mobilize its enormous coal reserves.

I do not want to give the impression, incidentally, that, other things being equal, a brief coal age might be an unreasonable cost for saving some of the world's larger forests. Economic theory cannot provide definite answers to puzzles of this kind, but it can suggest some meaningful questions that must be raised. In addition to destroying sinks for carbon dioxide, deforestation at one place can have important effects on soil erosion and rainfall at many other places. For example, many scientists are virtually certain that the deforestation taking place in the Amazon Basin will eventually have serious implications for the entire global weather system. Thus, assuming that the sole utility of a forest is derived from its timber and calculating its social value on this basis results in obtaining only a lower bound to the actual social value of the forest. To obtain a meaningful estimate of the social value of total or partial deforestation, all future costs resulting from deforestation must be added to this lower bound. Unfortunately, most of these future costs cannot be determined at the present time, although it cannot be excluded that some very unpleasant experiences or expenses might eventually be visited upon the industrial countries because of the disappearance of large expanses of vegetation around the world.

The Immediate Threat: Sulfur Dioxide

This section is divided into two parts. First, the more mundane aspects of SO_2 emission and control will be treated; and then there will be a short mathematical discussion of pollution abatement that underlines atmospheric pollution as a stock rather than a flow problem. Needless to say, these technical materials are secondary to the main presentation, since as is usually the case they are merely a restatement of several arguments that can be found elsewhere in these pages.

When fossil fuels are burned, gasified, or liquefied, they are transformed in one degree or another into their component parts: water, carbon, sulfur, arsenic, zinc, lead, mercury, antimony, copper, manganese, magnesium, gold, and beryllium. Until a few decades ago the earth's atmosphere was generally considered sufficiently capacious to render most of these substances harmless by simply reducing their concentration, in a given airspace, to an insignificant amount. As things have worked out, however, the growth in the capacity to pollute has increased by such an amount that this assumption is generally invalid.

The chief polluting agent among the fossil fuels is coal, which contributes more than half of the man-made sulfur emissions, and 80 percent of the sulfur emitted from stationary fossil fuel combustion. It has been estimated that in the United States, a 1,000-MW coal-fired electricity generation plant using 10,000 tons of coal per day would cause an annual emission of 8 million tons of carbon dioxide, 50,000 tons of sulfur dioxide, 20,000 tons of nitrous oxide, and between 25,000 and 250,000 tons of particulate matter, depending on how thoroughly the coal is washed before it is used and chimney gases are cleaned before being released.

But such things as the cleaning of coal or the scrubbing of stack gases (that is, flue-gas desulfurization) are not panaceas. Cleaning means first crushing and then washing coal, and finally disposing of consideragle waste—including some coal; but chemical bonding limits the removal of sulfur to less than 20 percent, and at present this operation can raise the price of coal by $5 to $10 per ton. Scrubbing, which is carried out by injecting materials such as limestone and water into a smokestack to react with the sulfur in exhaust gases, is also costly, and in addition makes it necessary to dispose of huge quantities of solid waste known as sludge. Because scrubbers are an add-on system, treating the problem after the fact rather than preventing it, they have no effect on the combustion process itself, and in some cases can prevent 90 percent or more of a plant's sulfur production from being released into the atmosphere. On the other hand, they can account for at least 10 percent of a plant's capital cost, and probably close to 20 percent if high sulfur coal is being used. Scrubbers must also be cleaned, which means that extra units should be installed if noninterrupted scrubbing is to take place. In addition, under certain conditions, scrubbing stack gases can produce sulfuric acid that must be collected and efficiently disposed of. (The scrubbing of stack gases at the Fulham and Battersea power stations in

London had to be abandoned because sulfuric acid was being washed into the Thames.)

This kind of situation has led many experts to insist that the best devices for controlling sulfur dioxide emissions are very high chimneys (that is, smokestacks) that eject gases very high above ground level. The dilemma here is that while high chimneys can often prevent excessive concentrations of sulfur residues from settling in the vicinity of a factory or power plant, they create major problems for distant locations. In fact, a large number of high stacks scattered over a major industrial area can theoretically lead to sulfur dioxide concentrations in distant regions that are a multiple of those which would exist locally under the worst possible conditions.

It has thus gradually become clear that the optimal solution to this problem involves burning coal more cleanly and efficiently in the first place. That leads us to so-called fluidized-bed combustion, which is reputed to be capable of reducing sulfur dioxide emissions by an average of 90 percent. It involves crushed coal being fed into a boiler compartment containing small particles of limestone placed on top of a metal grid. When hot air is forced up through the grid the coal and limestone mixture begins to float and expand, and eventually the solid particles commence displaying the properties of a boiling liquid—the origin of the expression "fluidized bed." At an operating temperature of 1550 degrees Fahrenheit the limestone particles chemically absorb sulfur and nitrous oxides from the burning coal, producing calcium sulphate (a benign solid waste) that is drained off through pipes at the bottom of the boiler. Thus a very large fraction of the pollutants are removed before the flue gases leave the boiler. Even high sulfur coal can, at least in theory, be successfully treated by this process, and it is also said that equipment of this type can burn a variety of solid wastes. Fluidized bed units are produced by some of the leading manufacturers of conventional boilers, including the Foster Wheeler Corporation, Babcock and Wilcox, and the Riley Stoker Corporation (a subsidiary of the Ashland Oil Company).

As bad luck would have it, though, fluidized bed installations are not yet commercially available in the sizes required for the larger electric power plants; and it is unlikely that they will appear in sufficient numbers for the remainder of this decade. For the most part, both commercial and pilot installations in the United States and Europe are comparatively small, although research in places like Germany is being speeded up by disclosures of the extensive damage caused German forests by pollutants originating in fossil fuels. German installations of up to 200 MW are under construction, and a demonstration plant at Esertal is already generating 129 MW. But sufficient problems have been encountered with equipment of this type (such as incomplete combustion, suboptimal sulfur removal, and clogging) to make it clear that first-generation units for very large installations may never be introduced on a commercial scale. The general feeling now is that combustion and desulfurization, which are two different processes, and at present take place in the same bed, must be physically separated, and that this can be done without sacrificing efficiency. Still another possibility is a circulating bed where fuel and limestone circulate throughout the reactor to extend the time (and thus the completeness) of combustion and desulfurization.

The combustion of coal also results in discharges of nitric oxide (NO) and nitrogen dioxide (NO_2). Nitrous oxides derived from coal account for 20 to 30 percent of those originating in man-made sources, which in turn amount to only 10 percent of the total from natural sources. The main problem here is one of concentration, particularly in urban and industrial areas with special atmospheric conditions. Coal also contains chlorine, which is converted to chloride compounds and hydrochloric acid gas (HCL). Chlorides are mostly found in ash, but some vaporize and cause severe corrosion problems in combustion equipment. Another product of coal is ash, the heaviest part of which (up to 40 percent of the total) remains in the combustion chamber. The lighter or fly ash mixes with the combustion gases and with the small quantities of particulate matter generated by the components of these gases. The present opinion in the United States, for example, is that these particulate emissions are dangerous to human health, and present regulations call for new coal-fired installations to possess equipment capable of capturing at least 99 percent of these particulates.

The use of coal requires large quantities of land for such things as storage and transportation, in addition to the premises of coal-burning installations; a great deal of land is also required for the disposal of waste products. According to the Environmental Protection Agency (EPA) the total land use requirements of a 1,000-MW powerplant over its lifetime may be as much as 1,000 acres. In particular, the land required for the disposal of ash and scrubber sludge, where the latter requires up to five times as much space as an equivalent weight of ash, may not be obtainable in the vicinity of coal-burning facilities at moderate prices. The Office of Technology Assessment (OTA) has estimated that more than 100,000 acres of land will be needed for ash and sludge disposal by the year 2000.

Given these many agonies, federal and local authorities in the United States have been moving toward an environmental strategy characterized by a no-effects threshold for pollution. Among other things, the gradual evolution of this kind of policy, backed up by political pressure, has led to a discernible cleanup of the air over many U.S. cities, and by the late

1970s a sizable improvement could be noticed in comparison with the situation existing in the mid-1960s. However, even with the atmosphere over many metropolitan areas visibly cleaner, questions remained concerning the toxicity of SO_2 and various particulates. Eventually it was postulated that a genuine danger to human health is present in the sulfate compounds that are formed when SO_2 mixes with the oxygen and trace substances that are found in polluted air spaces. Typical of these sulfates is sulfuric acid, and sulfates of zinc and iron, which materialize as ultrafine droplets of solid particles with a tendency to remain suspended in the atmosphere. Experimental evidence implies that sulfates of this nature might be many times more irritating to the respiratory system of animals and humans than oxides of nitrogen, hydrocarbons, and SO_2 plus particulates. Some scientists insist that most of the damage from pollution comes from sulfates; and the same position has been taken by economists like Lester Lave and Eugene Seskin.

The upshot of all this is that emissions of SO_2 from the main coal-burning regions of the world are being chemically converted to such things as sulfate in the atmosphere, carried by the prevailing winds for hundreds or thousands of miles, and then incorporated into the precipitation in the form of sulfuric acid. (A similar process produces nitric acid.) This is the origin of acid rain, a plague that has been particularly noticeable in Scandinavia, the northeastern United States, and eastern Canada. Moreover, the principal problem appears to be the speed with which intolerable concentrations of potentially dangerous substances tend to build up in the atmosphere. Thus, of late, politicans and civil servants have tended to ignore ivory tower academic approaches to dealing with this threat, and instead have chosen a direct attack on the SO_2 problem. In the United States, direct attack can be translated into the Waxman-Sikorski bill, which is now considered almost certain to pass the Congress in one form or another, and among other things requires the reduction of SO_2 emissions across the country by 10 million tons per year. Flue gas desulfurization systems are being made compulsory for the fifty largest coal-burning power plants in the United States, which should eliminate about 5 million tons of SO_2 per year, while eliminating the remaining 5 million tons will be left to negotiations between the EPA and the governors of the various states.

Admittedly, there will be some difficulty in financing these pollution control facilities in a depressed economy. Sales of plant flue gas scrubbers expanded very rapidly until 1980, when they were a victim of the recession brought on by the second oil price shock. U.S. sales of flue gas desulfurization equipment rose from $7.3 million in 1976 to $665 million in 1980. In 1981 sales fell to $185 million. If the Waxman-Sikorski bill is passed in its present form, scrubbers will be required for 158 generating units, at a total cost of about $5 billion. The question then becomes whether this charge should be paid by the utilities, by their customers in the form of higher prices for electricity, or by the general population.

As far as I am concerned, since acid rain and other pollution problems can seriously disrupt health and economic life, both in the present and in the future, the same kind of equity considerations apply here as to investment in defense and education. The cost of this equipment, while comparatively small if spread over the entire population—or for that matter just the population of those regions of the United States most severely affected by acid rain—is far from negligible if it must be supported by a relatively limited subset of the population. And if by some chance this burden fell primarily on utilities, it could tend to exacerbate macroeconomic difficulties that are already at the critical stage in many parts of the United States. The thing that the reader should pay particular attention to here, however, is the comparatively small amount of money that would be required to reduce sulfur emissions to a negligible level. Probably $10 billion for five years so so, and then perhaps one-half of this amount for another five or ten years.

A Digression on Stock and Flow Effects

Since it has been implied that a large part of environmental destruction can be attributed to the eventual buildup of undesirable emissions into a stock of pollution, some readers may find the following elementary mathematical demonstration interesting. To keep things simple, I assume a single industry producing some kind of composite good, using an amount R_1 of a composite resource. This industry emits an amount of pollution that is proportional to the output of the good that is being produced, and is equal to $a_{21}x_1$, where x_1 is the gross output of the industry. Now, by assuming away intermediate goods, gross output can be set equal to net output, which will be called y_1.

There is also an industry producing an antipollutant, x_2. This industry features an input of the single factor of production that is equal to R_2, and so the total use of this particular resource is $R = R_1 + R_2$. Finally, there is a certain tolerable level of pollution (per time period) that can be called y_2. Given this information, the following equation system should be self-explanatory:

$$x_1 = y_1 \tag{6.1}$$

$$a_{21}x_1 - x_2 = y_2 \tag{6.2}$$

$$R = R_1 + R_2 = a_{31}x_1 + a_{32}x_2 \tag{6.3}$$

Using equations 6.1 and 6.2, equation 6.3 can be written as:

$$R = (a_{31} + a_{32}a_{21})y_1 - a_{32}y_2 \tag{6.4}$$

A kind of budget constraint or production possibility curve can now be constructed using equation 6.4, but first it should be recognized that if it is decided not to suppress any pollution, all of the given resource R will go into the production of y_1, and from equation 6.3, $\bar{y}_1 = R/a_{31}$. Similarly, if it is decided not to permit any pollution at all, then $y_2 = 0$, and, from equation 6.4, $\hat{y}_1 = R/(a_{31} + a_{32}a_{21})$. Thus:

$$y_1 = \frac{R + a_{32}y_2}{a_{31} + a_{32}a_{31}} \tag{6.5}$$

With:

$$\frac{R}{a_{31} + a_{32}a_{21}} \leq y_1 \leq \frac{R}{a_{31}}$$

The assumption will now be made that the utility of this community is a function of its access to y_1, and the flow of pollution—y_2—experienced each period as a result of producing y_1. We thus have $U = U(y_1, y_2)$, and with y_1 a "good" and y_2 a "bad," we also have $\delta U/\delta y_1 > 0$ and $\delta U/\delta y_2 < 0$. Accordingly, if we have an interior solution, we can show the solution to our problem of choosing y_1 and y_2 as a simple tangency between the budget constraint and one of the utility curves in the preference system.

As shown in figure 6–2, utility maximization implies y_1 units of the commodity and y_2 units of pollution per period. But as emphasized previously, much of the damage done by pollution is the result of its building up into intolerable concentrations over time. Thus we need to transform the flow of pollution y_2 into a stock of pollution. This can be done by taking into consideration both the increment of new pollution y_2 and the capacity—per period—of the atmosphere to dissipate the existing stock of pollution S. With k the pollution depreciation coefficient, kS of pollution is dissipated per period, and so we have:

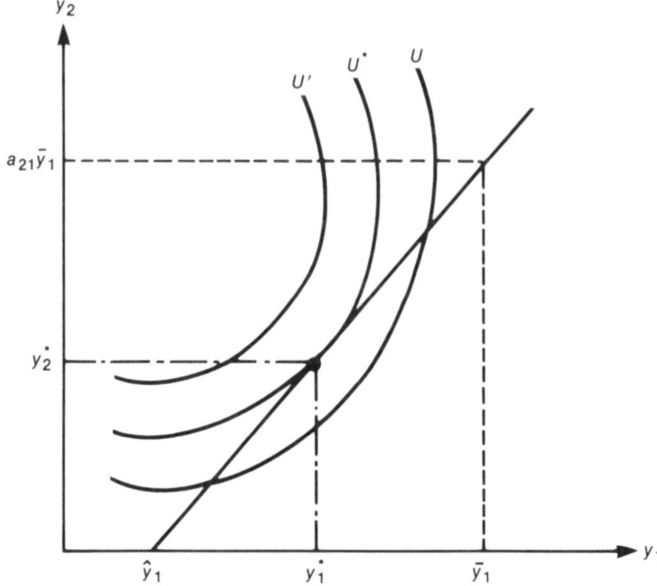

Figure 6–2. Utility Maximization with One Commodity and Pollution

$$\frac{dS}{dt} = (a_{21}x_1 - x_2) - kS = y_2 - kS \tag{6.6}$$

This simple first-order differential equation can be easily solved to give:

$$S(t) = \left[S(0) - \frac{y_2}{k} \right] e^{-kt} + \frac{y_2}{k} \tag{6.7}$$

From this expression it is easy to see that $S(t) \rightarrow y_2/k$ when $t \rightarrow \infty$. Although this may seem interesting from the strictly algebraic point of view, it does not represent the arrangement portrayed earlier in this discussion, where environmental damage—at least in theory—had the capacity to reach virtually any finite level. What we have above is an analogy of the tortoise and hare parable: the larger the amount of pollution, the larger the amount dissipated. Thus, unless $k \rightarrow 0$, the stock of pollution can never become infinite.

The theoretical way to get a more realistic result is to let k be a function of S, with $dk/dS < 0$; but the resulting differential equation would not be easy to solve or interpret. I therefore employ the following simple difference equation to bring out this point:

$$\begin{aligned} S_t &= S_{t-1} + y_{2t} - k(S_{t-1} + y_{2t}) \\ &= S_{t-1} + y_{2t} - (k_o - \theta S_t)(S_{t-1} + y_{2t}) \end{aligned} \tag{6.8}$$

This is almost an analogy of equation 8.6, except that the dissipative power of the environment (that is, atmosphere) can be overwhelmed by a sufficiently large stock of pollution. Numerical experiments with this expression should convince the reader immediately that S_t can go to infinity if θ and y_2 are sufficiently large with respect to k_o.

Finally, a situation can obviously exist where y_2 increases over time. Then we might have $y_2(t) = y_2(0)e^{nt}$. This relation can be put in equation 6.6, yielding a differential equation with the solution:

$$S(t) = e^{-\int k dt}\,[A + \int y_2(0)e^{nt}e^{\int k dt}dt]$$

Where A is an arbitrary constant that can be determined using the initial conditions $S(t) = S(0)$ when $t = 0$. The integration of this expression is perfectly straightforward and yields at once:

$$S(t) = \left[S(0) - \frac{y_2(0)}{n + k}\right]e^{-kt} + \frac{y_2(0)e^{nt}}{n + k} \tag{6.9}$$

We see immediately that when $n = 0$ we get equation 6.7, but when $n > 0$ the stock of pollution continues to grow over time. Everything considered, however, this solution is far from ideal, and probably some nonlinear refinements should be introduced. There is no point in going into what these should or could be in this introductory analysis; but clearly a good start might be to let k become a function of S once more.

The Distant Threat: Carbon Dioxide

In the past 120 years the level of carbon dioxide (CO_2) in the air has risen by at least 15 percent to a concentration of about 350 parts per million (ppm). Assuming that present energy use trends are maintained, this could mean that the atmospheric concentration of CO_2 would reach 380 ppm by the year 2000, which implies a doubling sometime in the next 75 to 100 years. Table 6–2 provides some further information about the production of CO_2.

It is not exactly certain in the scientific sense what this increased CO_2 production will mean for the environment, but there is no reason to be optimistic. First of all, radical adjustments in the climate are theoretically possible as a result of the so-called greenhouse effect. An increase in the atmospheric concentration of carbon dioxide does not materially alter the amount of the sun's radiation reaching the surface of the earth, but there is a significant reduction in the amount of heat from the earth's surface that is radiated out to space. As a result a palpable warming of the earth's surface becomes very likely. Several years ago a number of scientists felt it possible to claim that even a doubling of atmospheric carbon dioxide would only increase global temperatures by about 0.25 degrees centigrade; but among others S. Manabe (1971) has presented a highly regarded calculation indicating that an increase of atmospheric CO_2 to 385 ppm would mean a temperatuure rise of 0.6 degrees centigrade, while Manabe and Wetherald (1975) have calculated that a doubling of the atmospheric concentration of CO_2 could lead to a temperature increase of three degrees. A considerable amount of uncertainty must, of course, be attached to these and other estimates; and it is also true that the presence of various greenhouse emissions, such as methane and chloroflurocarbons can affect the rate at which the temperature increases, as well as triggering feedback mechanisms that provide more cloud and rain and thus help to lower temperatures, but as far as I

Table 6–2
Areas Producing Carbon Dioxide and the Carbon Release of Some Important Energy Media

Areas	*1974 (%)*	*2025 (%)*	*Fuel*	*Carbon*
North America	29	9	Oil	19.2
OECD (minus North America)	25	11	Gas	13.7
Eastern Europe	25	31	Gas	23.8
Third World (minus China)	13	33	Solar Nuclear	0.0 0.0
China	8	16	Hydro	0.0
Total (billions of tons)	16	92		

Source: Institute for Energy Analysis.

can tell the three-degree estimate of a temperature increase is becoming increasingly visible in the scientific literature. (For example, it is found in the work of both the EPA and the American Academy of Sciences). However, for the record, it should be noted that even small changes in the average temperature can have provocative consequences. Around the year 1300, with the temperature in the Northern Hemisphere at about the same average level as today, a fall in the average temperature of around one degree was enough to make ice skating on the Thames possible during a few weeks of the winter. Similarly, a one-degree average rise in the temperature might cause a disagreeable increase in the amount of bacteria in the surroundings.

Without going too deeply into the matter, let me make it clear that for a civilization that is as fine-tuned to the present climate as is ours, an average increase in the global temperature of three degrees could be serious. An average annual increase of three degrees could mean a rise of up to ten degrees in the polar regions during certain periods of the year, while at the same time it could happen that there is no change at all in the vicinity of the equator. But it is the pole-to-equator temperature difference that determines wind patterns around the globe; and these in turn determine how much rain falls, as well as where it falls. Changing pole-to-equator temperatures by ten degrees or more could lead to a geographical reallocation of agriculturally productive land that, together with incidental changes in productivity, would result in a lower global agricultural output. An outcome that might be unfortunate for the entire world would be a reduction in the highly productive agricultural surplus areas of the United States. (Some theoretical evidence seems to suggest that, with such temperature changes, Canada and the Soviet Union will be relatively favored, while the United States will be a net loser.) It has also been judged plausible that increasing CO_2 levels could unleash substantial amplification (or multiplier) effects. Increasing soil temperatures might escalate the production of CO_2, while higher water temperatures could conceivably decrease the capacity of the oceans to function as a sink or repository for CO_2 because the decreasing alkalinity of saline seawater will decrease the solubility of carbon dioxide at or near the surface of the water.

An interesting attempt to come to grips with the economics of this topic has been made by Edmonds and Reilly (1983a, 1983b, 1983c), while the general reader will find Woodwell (1978) useful. Nordhaus (1982) has approached the carbon dioxide impasse as a complex problem in welfare economics that, at least in theory, can be solved if people of good will are prepared to put their trust in some elementary results from the calculus of variations—or its modern version, optimal control theory. On the other hand, Bohm (1979) has contributed a helpful but elementary discussion of ozone protection policies where the international issues are similar, if not identical, to those associated with CO_2. The remainder of this section draws heavily on these expositions.

One of the basic results obtained by Edmonds and Reilly is that CO_2 growth will be relatively slow during the remainder of this century, but then will accelerate. As it happens, this judgment follows almost directly from their assumptions about the profile of fossil fuel discovery and use. If, as shown in figure 6–3A, the production of fossil fuels takes on the characteristics of a logistic function, and CO_2 production is a direct function of fossil fuel production (and use)—which to a considerable extent it is—then by making the additional assumption that large bodies of water (such as oceans) and organic materials (such as vegetation) continue to function as sinks that absorb and to some extent neutralize CO_2, we get the CO_2 concentration curve shown in figure 6–3B.

The values shown in figure 6–3 are indicative rather than precise and roughly correspond to those given in Devins (1978). To confirm the patterns shown in figure 6–3B, let us remember that logistic functions are usually put in the following form:

$$Q_c = \frac{Q_m}{1 + be^{-at}}$$

where Q_c is cumulative output, Q_m is maximum output, and a and b are parameters. This expression can be modified to take into consideration the ability of organic materials and the oceans to dissipate the prevailing stock of CO_2. Thus we get:

$$Q_c = \frac{Q_m e^{-\phi t}}{1 + be^{-at}} \tag{6.10}$$

Here ϕ is a dissipation or depreciation rate acting on the stock of CO_2 in the atmosphere. It is a comparatively simple matter to estimate values for a, b, and ϕ that will give production and concentration curves similar to those presented in figure 6–3. The production of CO_2 can be obtained by differentiating the basic logistic function, or:

$$\frac{d}{dt}Q_c = \frac{d}{dt}\left[\frac{Q_m}{1 + be^{-at}}\right] = \frac{Q_m}{1 + be^{-at}(2 + be^{-at})} \tag{6.11}$$

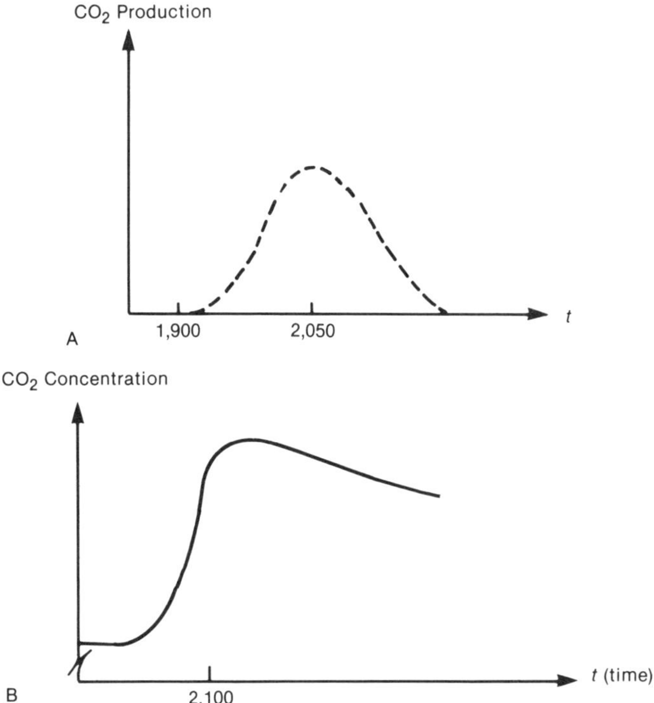

Figure 6–3. CO_2 Production and Concentration (assuming that CO_2 production is a direct function of the use of fossil fuels).

Now we can go to a theme that, I am sorry to say, may be well beyond the scope of mainstream economics, at least in its details. This is the international management of CO_2 production. What we have here is a situation where the accumulation, over time, of CO_2 in the entire upper atmosphere is a function of the measures taken by independent national governments to minimize the local release of CO_2. There is, ironically enough, at least one bright spot in this melodrama due to the fact that the burning of fossil fuels can cause nearby damage in the immediate future, as well as hypothetical damage in the distant future. The Waxman-Sikorski bill, for instance, is an instrument for the suppression of acid rain, and because it will make the burning of coal more expensive, will probably lead to an aggregate decrease in the use of carbon-rich fuels.

But, clearly, some thought must be given to a straightforward attack on the carbon dioxide predicament. William Nordhaus, a member of the carbon dioxide assessment committee of the National Research Council, has suggested (or "concluded," to use his terminology) that a warming of the atmosphere will have a major impact on agriculture and coastlines, as well as global GNP (which he pictures as declining by from 5 to 12 percent). All this implies a lethal threat to our environmental and perhaps ecological heritage, and in the words of the brilliant jurist Benjamin Cardozo, "danger invites rescue." The only question is how.

Certainly, as Peter Bohm has indicated, one country (for example, the United States) cannot become the gendarme of the upper atmosphere and place a large part of the burden of global environmental protection on its own population without getting something in return, although it is important for countries like the United States to make a strong commitment to environmental sanity to influence its allies and wellwishers—and even more important for countries like Sweden. To my way of thinking, innovations like the Waxman-Sikorski bill are an important first step in this process, since if U.S. pollution can be reduced in a fairly short time, perhaps influential politicians and economists elsewhere might forget about algebraically elegant but impotent tax schemes and lend their support to regulations calling for an immediate quantitative restriction on dangerous emissions. (Baumol and Oates (1971) have written that "Experience would soon permit the authorities to estimate the tax levels appropriate for the achievement of a target reduction in pollution." In terms of the real world, this statement has no basis in fact; however, as theoretical contrivance it can probably be granted a certain naive credibility or excellence.)

As indicated previously, the basic quandary we are facing is that the earth's atmosphere is a common property resource, and its exploitation by one person or one country can cause external benefits or disbenefits to others. Moreover, the alternative exists for some individuals, organizations, or countries to take advantage of the restraint or altruism of others. For instance, if the regent of Monaco became seriously worried about the CO_2 problem and ordered the burning of 1,000-franc notes in the furnace of the casino instead of coal, news of the impending passage of the Waxman-Sikorski bill might result in His Majesty finding another use for this picturesque lucre. Furthermore, by phrasing the conundrum in this manner, we immediately see that it can be given a more detailed inspection within the schema of a particularly useful contrivance of elementary game theory, the so-called prisoner's dilemma.

Consider the following tableaux, which represent the matrix form of a two-person game. There are four possible outcomes, with each outcome depending on actions of the two players, who can be referred to as Sam and Ivan. In the matrix on the left I have inserted some arbitrary numerical outcomes that the reader can think of as social profits. The first number refers to Sam and the second to Ivan. Each has two strategies involving the production of CO_2: x (no change, or status quo), and y (restraint). The righthand tableau corresponds to that on the left only, and for pedagogical reasons, the numerical outcomes are transformed into utilities. Notice that in the arguments of these functions, the first term refers to the strategy of Sam, and the second to that of Ivan.

		Ivan x	y
Sam	x	100/100	200/50
	y	50/200	150/150

	x	y
x	$U_I(x,x)$ $U_S(x,x)$	$U_I(x,y)$ $U_S(x,y)$
y	$U_I(y,x)$ $U_S(y,x)$	$U_I(y,y)$ $U_S(y,y)$

Making the usual assumptions of completeness, transivity, and reflexivity about Sam and Ivan's behavior means that they would rank these outcomes as follows:

$$\text{Sam: } U_S(x,y) > U_S(y,y) > U_S(x,x) > U_S(y,x)$$

$$\text{Ivan: } U_I(y,x) > U_I(y,y) > U_I(x,x) > U_I(x,y)$$

Here it should be obvious that, for Sam, x is the dominant strategy: if Sam chooses x and Ivan y, Sam achieves his best outcome. By way of contrast, if Sam chooses y and Ivan x, the outcome for Sam is the worst possible. Since the game is symmetrical, the same is true for Ivan—which is apparent from the second row of inequalities. Since the condition of the game is that they were both playing x to begin with, it is clear that in a noncooperative situation they continue to do so. x is therefore the equilibrium strategy for both players, and is generally known as a Nash equilibrium or a Cournot-Nash equilibrium. The numerical value of the outcome in this case is 100/100.

On the other hand, if agreements are enforceable, clearly the outcome 150/150 is preferred by both players to the noncooperative outcome 100/100. The cooperative outcome also happens to be Pareto-optimal, in that once this outcome is attained, a movement to any other outcome makes either Sam or Ivan worse off. However this does not imply that, once attained, players would be willing to accept the Pareto-optimal arrangement if they were perfectly free to move. If one or both of them felt that a tremendous gain could be realized by playing x instead of y, or feared that by playing x a huge loss would result if the other changed his strategy and played *y*, the temptation for both of them to play *y* would be irresistible.

The next question then concerns getting both players to agree to adhere to the Pareto-optimal outcome—that is to say, to cooperate. If they both know game theory this might be easy, because most books on that subject tend to stress that whenever cooperation is possible, then almost without exception it is in the interest of both players to see that it takes place. Moreover, most real life conflict or potential conflict situations are not games but supergames (that is, repeated games), where there is a possibility for rational players to learn from their mistakes and alter their behavior. One of the main problems that we are facing here, however, is that the players may learn too late. For instance, as dangerous as the present arms race is, I personally expect both sides to come to their senses eventually and scale down their spending on weapons, and perhaps also the existing level of armaments. Scaling down the stock of CO_2 in the atmosphere after it reaches a critical level, however, might take hundreds of years, and presumes an almost unrealistic flexibility and generosity on the part of the main producers of CO_2.

There is also the possibility of one player attempting to directly influence the other(s). Lester Lave (1965) has found that in two-person repeated games played under laboratory conditions, cooperation can often come about if one of the players makes a point of setting a good example. It seems to me that it is always good business to attempt to obtain a moral advantage over competitors or colleagues by setting the right example; but I hardly see the advantage in pretending that laboratory tournaments for pennies and affection are equivalent to full-fledged operations at the Mirabel, Cafe Boulevard, or Palais des Nations. Without going too deeply into the matter, let me say that the best way to deal with this dilemma is to make it part of a general program to reduce international tensions, transfer at least some resources from military to civilian uses, and engage in activities can could be interpreted as idealistic by both blocs and by third parties. Not much imagination is required to see that an all-out war against atmospheric and marine pollution would be perfect in this respect.

At least one important topic is missing from this discussion: intertemporal allocation and efficiency. For some time now William Nordhaus (1982) has supported the obsession that a carbon tax is the right way to accelerate the transition from fossil fuels. He has even made some calculations of suitable values of this tax through the middle of the next century, under alternative assumptions about such things as discount rates and economic growth. However, as far as I know, the notion of a superabundant source of energy provided by a backstop technology is as fresh today in Professor Nordhaus's mind and his research as it was when he first elaborated it, and frankly I doubt whether it has had a positive influence on his prescription for eliminating CO_2. It seems likely that the cost advantages of using fossil fuels are such that even if the tax rates calculated by Nordhaus were applied in full, on a worldwide basis, there would only be a marginal effect on the build-up of atmospheric CO_2.

Concluding Remarks

I will conclude this chapter with a few comments on some environmentally friendly technologies that also involve the use of coal. As mentioned earlier, fluidized bed combustion is an important technology—but it will require several years before it can be used in the largest coal-burning installations. Another promising technique is the PFBC process developed by ASEA of Sweden. This process releases only one-fifth of the maximum emissions of sulfur dioxide permitted by the Swedish Department of the Environment. The only question remaining is whether it can be used in other than pilot-sized installations.

It has also been said that the use of coal-water mixtures will bring about an improvement in the environment. A fuel is blended of 70 percent coal, 29 percent water, and 1 percent proprietary additives, and used as a liquid in oil-burning installations. As a result there is no need for costly storage, coal-handling, and coal-burning equipment. Apparently there are still some important technical problems to overcome before this technique can be employed on a large scale, but it has the endorsement of some of the most important research institutes in the United States. Let me also remind the reader that the Japanese have proved singularly successful in reducing pollution. During a ten-year period, they have reduced sulfur oxide emissions to one-quarter of the 1970 amounts. The secret of their success was in the *enforced* installation of equipment for the desulfurization of smoke gases. In that country it has been estimated that environmental destruction from poisonous fumes is now under control.

Finally, it should be noted that gas can be burned with coal. It may, in fact, be possible to burn some combination of gas and coal without scrubbers, and still maintain air quality. In dual-firing systems, gas and coal could be burned separately or together; but in any event the intention would be to have gas contribute 15 to 30 percent of the required energy input.

Appendix 6A
Pollution Taxes and Intertemporal Utility in the Presence of Pollution

Pollution Taxes

A few short remarks are in order here concerning pollution taxes. To begin, in theory, it is hard to object to them. Much governmental regulation is too inflexible, and allocations through the price system—which take into consideration individual circumstances—are often preferable. Furthermore, when intervention is necessary, the government can use pricelike mechanisms. Obnoxious emissions can, at least in theory, be taxed at rates reflecting the costs imposed on others and thus eliminated completely.

To show this, let us take T as profits (exclusive of pollution-based expenses), E emissions, E' the maximum level of emissions, t the pollution tax rate, and ϕ the cost of eliminating a unit of pollution. Thus we get:

$T - tE' = \overline{T}_1$ Total profit if no pollution is eliminated
$T - \phi E' = \overline{T}_2$ Total profit if emissions completely eliminated

For any in-between amount of emissions we must have:

$$0 \leqslant E \leqslant E' \quad \text{and so} \quad \overline{T} = T - tE - \phi(E' - E)$$

From this last expression we can write:

$$\frac{d\overline{T}}{dE} = -t + \phi$$

This makes it clear that:

$$\text{if} \quad t > \phi \quad \text{then} \quad \frac{d\overline{T}}{dE} < 0$$

and all emissions will be eliminated.

Intertemporal Utility

The problem dealt with by Nordhaus (1982) involves maximizing an aggregate utility function over time. For our purposes this function can be said to contain income and the stock of pollution as arguments. It is simple to deal with this matter in terms of optimal control theory, but in fact greater insights are possible if the calculus of variations is used. Using the same variable employed in chapter 6, we are interested in maximizing the following expression:

$$\int_{t_o}^{t_f} U(Y, S)e^{-rt}dt \quad with \quad \dot{S} \equiv \frac{dS}{dt} = \alpha Y - kS$$

The constraint can obviously be written:

$$Y = \frac{\dot{S} + kS}{\alpha} = \theta_1\dot{S} + \theta_2 S \quad \text{where} \quad \theta_1 = \frac{1}{\alpha}, \theta_2 = \frac{k}{\alpha}$$

Accordingly, the integral to be maximized is:

$$\int_{t_o}^{t_f} U(\theta_1 S + \theta_2 S, S)e^{-rt}dt$$

We can now apply the Euler-Lagrange conditions to this integral. First of all, the following calculations are necessary:

$$\frac{\delta U}{\delta S} = \left[\frac{\delta U}{\delta Y}\frac{dY}{dS} + \frac{\delta U}{\delta S}\frac{dS}{dS}\right]e^{-rt} = \left[\frac{\delta U}{\delta Y}\theta_2 + \frac{\delta U}{\delta S}\right]e^{-rt}$$

$$\frac{\delta U}{\delta \dot{S}} = \left[\frac{\delta U}{\delta Y}\frac{dY}{d\dot{S}}\right]e^{-rt} = \left(\frac{\delta U}{\delta Y}\theta_1\right)e^{-rt}$$

The last expression should now be differentiated with respect to t. This gives:

$$\frac{d}{dt}\left[\left(\frac{\delta U}{\delta Y}\theta_1\right)e^{-rt}\right] = e^{-rt}\frac{d}{dt}\left(\frac{\delta U}{\delta Y}\theta_1\right) - re^{-rt}\frac{\delta U}{\delta Y}\theta_1$$

As usual, the solution we are interested in is the steady-state solution. This implies that:

$$\frac{d}{dt}\left(\frac{\delta U}{\delta Y}\right) = 0$$

Thus, from the previous expression we get immediately:

$$\frac{\delta U}{\delta Y} = \frac{-(\delta U/\delta S)}{\theta_2 + r\theta_1} \equiv \frac{\alpha(-\delta I/\delta S)}{k + r}$$

What this says, as usual, is that a steady-state optimum features marginal benefit equal to marginal cost, where the latter is specifically defined as the disbenefit from an extra unit of pollution.

7
Electricity and Economics: An Introductory Survey

Thanks to OPEC, many economists have come to know a great deal about crude oil. Gas, coal, and nuclear energy are also discussed with considerable authority at the various energy meetings and symposiums that increasing numbers of economists and decisionmakers in the energy world are now attending. By way of contrast, electricity economics is a subject that even many people calling themselves energy specialists seem to be totally unfamiliar with; and it indeed plays only a minor role in most of the book-length literature on energy economics.

This being the case, I would like to offer an elementary presentation of electricity economics, and in particular to discuss some topics alluded to in my book, *The Political Economy of Oil*. The emphasis here will be on simplicity and the clarification of basic concepts; however, comprehending the difference between energy and power involves becoming fluent in the use of a few terms from elementary physics. The small amount of mathematics employed in the discussion is mostly of the junior high–secondary school variety; and when anything more complicated is used, it is separated from the main argument and put in subsections that can be skipped by the general reader.

In the most comprehensive possible sense, energy can be defined as anything that makes it possible to do work (that is, bring about movement against resistance). Energy takes many forms, and one of its most interesting characteristics is that all forms of motion, all physical processes, involve to one degree or another the conversion of energy from one state to another. For example, the chemical energy that is found in coal can be converted to active heat, which in combination with water will generate steam in a boiler. This steam can then be used to drive a turbine that, in turn, rotates the shaft of an electric generator, producing electricity. Note also that the rotating shaft implies the ability to do

---

I would like to thank the students in my course in energy economics at the IGS, University of Stockholm, for their stimulating comments on the materials in this chapter, as well as my colleagues at the University of Uppsala and Melbourne University.

physical work. For instance, it could be used to turn a merry-go-round or pull a cable car up a steep incline (although, admittedly, these may be very uneconomical applications). Analogously, the chemical energy in food can be transformed into mechanical energy—that is, the ability to do physical work or, for that matter, the ability to do mental work.

Power is the *time rate* at which energy is converted. The unit of power in the International System of Units (SI) is the watt, although the watt is so small that the kilowatt (1,000 watts) is usually employed. We shall be using this term extensively in this chapter and in order to put it in its proper perspective, the reader should consider the following example. An average pound of coal contains 12,500 Btu of energy, where a Btu is the quantity of energy in heat form needed to raise the temperature of 1 pound of water (which is approximately one pint) by 1 degree Fahrenheit. An average metric ton of coal (2,205 pounds or 1 tonne) thus contains 12,500 × 2,205 = 27,500,000 Btu. Now let us introduce two light bulbs. One of these produces a great deal of illumination, and has a power rating of 500 watts; the other is considerably weaker, and has a rating of only 50 watts. If the heat energy in coal is totally and perfectly transformed into electrical energy (that is, with 100 percent efficiency), 3,412 Btu are required to generate a kilowatt-hour (kWh) of electrical energy (where kWh is the unit in which electrical *energy,* as distinguished from power, is measured). The power rating of the bulbs—500 and 50 watts, respectively—informs us of the rate at which the energy potential of the coal is consumed; and so if the 27,500,000 Btu in a metric ton of coal is transformed into electricity in a perfect system, it could provide exactly 27,500,000/3,412 = 8,060 kWh of electrical energy. In other words, in a perfect system, the stronger of the two bulbs, which consumes power at the *rate* of 500 watts (0.5 kilowatt), could function for 8,060/0.5 = 16,120 hours. The other bulb would require 8,060/0.05 = 161,200 hours to consume the energy in a tonne of coal. In reality, the efficiency with which fossil fuel energy can be converted to electrical energy is well under 100 percent. An efficiency of about 32 percent seems to be typical for most of the United States, and so on the average it would require 10,500 Btu (3,412/0.32) to obtain 1 kWh of electricity.

As simple as all this seems, many readers accustomed to economic terminology may feel that something is wrong, somewhere. The problem, most likely, is that while electrical power is defined as a "rate," it is not associated with a time dimension. The only thing I can suggest here is to mull over the well-known analogy between energy and power, and distance and speed. Speed is the distance traveled per unit of time, while power is defined as the energy converted per time unit. By the same token, given a certain speed, distance is a function of the time spent traveling; although with an apparatus having a certain power rating, the total energy converted is a direct function of the time the apparatus is in operation. In most situations, our principal concern is likely to be directed toward power ratings. It is the horsepower (that is, power) rating of an automobile engine that determines its speed, and the power rating of a bulb that determines how much light it gives off. On the other hand, many of us are uninterested in the distance our automobiles have gone or will go; and since leaving engineering school I have never encountered anyone with the slightest curiosity about the operating lives of light bulbs—although I am sure that such persons exist.

In conjunction with these observations, the key thing to visualize now is that the large bulb in the previous example is milking the tonne of coal of its energy potential more rapidly than the small bulb. If we ignore the purchase price of the two bulbs, it should be clear that it is more expensive to use the larger bulb, because in each period it extracts a larger amount of energy from the tonne of coal than the smaller bulb. Put another way, the larger bulb is doing more work than the smaller bulb, since it provides more illumination per unit of time; and in certain situations this justifies its higher operating (and acquisition) cost. Remember too that, excluding the cost of metering, residential electric bills are usually proportional to the amount of energy used. In fact, if you examine a typical bill, you will almost always find the number of kilowatt-hours of electrical energy consumed during a given period clearly designated; and the amount that must be paid the organization providing the electricity is some fixed charge plus the number of kWh consumed multiplied by some tariff or charge per kilowatt-hour.

One further point should be made before we leave this introduction. In this discussion I have used the Btu as the basic unit for heat energy. This is the common practice in both the United States and Australia, with the *therm* (1 million Btu) being an important unit in Australia, and the quad (10^{15} Btu) occupying a similar position in the United States. In Sweden, however, the basic unit is the joule; and it seems that in serious scientific usage the joule is the preferred unit almost everywhere. Fortunately, it is a simple matter to go from one unit to the other: 1 billion joules (1 GJ) is equal to 947.8×10^3 Btu or, conversely, 1 Btu = 1.055×10^{-6} GJ. The usefulness of the joule for scientific work becomes apparent when we discuss units of electrical energy, where 1 watt for 1 second is equal to 1 joule.

The Fuel Cost of Electricity

Electricity is both a primary and secondary *fuel*, or technically useful energy. The most abundant primary fuels are coal, oil, natural gas, and electricity having a nuclear or hydro origin. Almost all the others used in industrial societies are secondary, being produced from primary energy sources. Here I am talking in particular about electricity generated in oil-, coal-, or gas-based power stations, as well as electricity-producing installations employing such things as town gas or coke. Table 7–1 gives some indication of the cost of electricity to industrial users throughout the world, as well as the distribution of electrical generating capacity by fuel type. Although the price of electricity is not always determined by the cost of the resources used in its generation, it generally tends to be the case that countries in possession of large amounts of energy materials, including dammable water, display comparatively low energy costs. This is especially true for Canada, the Republic of South Africa, Australia, and until recently the United States. The only major exception that I can think of at the present time is Britain, where the current government has—for reasons unknown to this economist—made it a point of honor to keep the price of electricity to British industry well above that being paid by its competitors. The disadvantages of not having ample domestic supplies of energy can sometimes be offset, however, by the timely and imaginative use of technology. The rapid expansion, after the energy price rises of 1973–1974, of nuclear generating capacity in Sweden and France is an example of this.

The major components of the cost of electricity are (1) the cost of fuel, (2) the cost of operations and maintenance, (3) the cost of transmission losses, and (4) the cost of capital. The cost of fuel and the fixed charges, or cost of capital, account for most of the cost of electricity, and these will be treated in some detail later. It can be emphasized here, in line with the observation at the end of the previous paragraph, that at least a limited trade-off is almost always possible between fuel and capital costs. For instance, in the case of Sweden and France, the average cost of generating electricity was decreased after 1973–1974 by reducing the amount of electricity generated in oil-burning installations—where equipment costs were fairly low, but fuel costs became extremely high—and increasing the quantity of electricity produced by nuclear facilities, where fuel costs were comparatively low, but capital costs high. In some countries, however, and particularly the United States at the present time, there does not seem to be much to gain by increasing the proportion of nuclear-based power in the energy picture. The problem appears to be that capital costs have become excessive due to a large increase in the time required to construct new installations and the greatly increased cost of insurance for nuclear installations.

The expression *efficiency* was used in the introduction when discussing the number of Btu necessary to obtain one kWh in an actual system. In practice, the term *heat rate* is employed, where heat rate can be

Table 7–1
Average Price Paid for Industrial Electricity and the Distribution of Electrical Generating Capacity in Fourteen Major Industrial Countries, 1980

	Electricity Price (Mills/kWh)	*Electrical Generating Capacity: 1000 MW*			
		Coal	*Oil-Gas*	*Nuclear*	*Hydro*
United States	40.5	243	199	55	73
Canada	18.7	15	13	6	48
Belgium	49.3	6	1	2	1
Denmark	49.7	4	3	—	—
France	35.7	17	9	17	11
West Germany	54.3	34	22	9	6
Italy	48.5	3	26	1.5	15.5
Luxembourg	37.7	—	—	—	—
Netherlands	48.3	2	13	1	—
Spain	34.1	6	9	1	14
United Kingdom	70.3	42	12	5	1
Japan	60.0	8	78	15	28
Australia	34.3	16	2	—	6
Republic of South Africa	25.1	16	—	—	1

Note: Totals are not given since a small amount of electricity is generated in these countries using other fuels. By definition, 1 mill = 0.1 cents. See the discussion in the text.

defined as 3,412 Btu/efficiency. Furthermore, knowing the cost of fuel in cents per million Btu (¢/MBtu), and the heat rate, will permit calculation of the cost of coal at the generating plant is $40 per short ton (2,000 pounds), which was the average landed price of steaming (that is, thermal) coal at Rotterdam in the second half of 1982. Again assuming 12,500 Btu/pound of coal, there are 25,000,000 Btu/short ton. The cost of 1 million Btu is therefore 40/25 = $1.6 = 160 ¢. Taking efficiency as 32.5 percent, the heat rate is 3,412/0.325 = 10,500 Btu, and the fuel cost of electricity is:

$$\frac{160\ \text{¢}}{1{,}000{,}000\ \text{Btu}} \times \frac{10{,}500\ \text{Btu}}{1\ \text{kWh}} = 1.68\ \text{¢/kWh}$$

In the electricity generating industry it is sometimes the custom to express energy costs per kilowatt-hour in terms of mills, where 1 mill = 0.1 ¢. Thus the previous figure becomes 16.8 mills/kWh. A brief examination of fuel and other variable costs of electricity generation in Sweden makes it clear that in existing installations, nuclear power has a decisive advantage over coal, oil, and gas; but it seems likely that the least expensive electricity can be obtained from hydroelectric installations, and this would also be true for new capacity.

The costs associated with such things as licensing, operations, and maintenance will not be taken up here, since these are straightforward expenses that can be immediately expressed in terms of kilowatt-hours. The same is true of fixed costs associated with transmitting electricity from generating facilities to the point of consumption, and also transmission losses. Although the latter can, in some situations, be quite large, they are generally held to a modest level by extremely high voltages that make national or even international grids feasible. For example, there is a 765,000-volt transmission line in Canada, and even higher voltages are being contemplated in the Soviet Union and the United States. As students of secondary school physics perhaps remember, for a given power, current is inversely proportional to voltage; and since transmission losses are proportional to the square of the current, increasing the voltage by a factor of two reduces losses by a factor of four.

Still, in a number of applications, it is a mistake to go through the process of using a fossil fuel to generate electricity, instead of making direct use of the fuel. As it happens, the conversion efficiency of coal to electricity may be much lower than the value given earlier, since that value did not consider the energy required to obtain coal. It has been suggested that for some countries—for example, England—the aggregate conversion efficiency of coal to electricity may be as low as 25 percent. There, for every kilowatt-hour of electricity produced, as much as 4 kWh of primary fuel input may be necessary. Clearly, for all its convenience and cleanliness, the price of electricity seems rather high when considered in these terms.

Capital Cost

Next we come to the matter of the capital cost, and what I will present here is a shortened version of the discussion given in *The Political Economy of Oil*. To begin, consider the plight of someone borrowing a sum of money. This person must commit himself to repaying this debt, generally in installments, which can be described in terms of two distinct components: *interest* charges and *amortization* charges. In some circumstances the amortization charge is referred to as a repayment of part of the principal, which is the amount borrowed. As will be made clear later, when we are dealing with a physical asset, the amortization charge can be equated to the *depreciation* charge, or the charge associated with wear and tear on the item. In addition, the sum of the yearly interest charge, and the amortization (or depreciation) charge, is called the *capital cost*. On the other hand, the *investment cost* is the price of a machine or structure.

To get a better idea of what this is all about, suppose that you are in a position to lend \$1,000. If the rate of interest is 10 percent, you will receive \$100 per year as interest income, every year, until the \$1,000 has been repaid. Unfortunately, the matter of amortization—or the repayment of the principal (\$1,000 dollars)—is not so simple, since it is necessary to specify the timing of repayments. For example, assuming that \$1,000 is lent for a period of two years, then only one amortization payment would mean that the lender would have \$1,210 at the end of two years. This could come about through the lender receiving an interest payment of \$100 at the end of the first year, which is put into a bank and transformed to $100(1 + r) = 100(1 + 0.1) = \$110$ at the end of the second year; and then receiving another interest payment, plus the principal of \$1,000 at the end of the second year. Now, at the end of the second year, the lender has $110 + 100 + 1{,}000 = \$1{,}210$. Another possibility is simply a lump sum payment at the end of two years of $1{,}000(1 + r)(1 + r) = 1{,}000(1 + r)^2 = 1{,}000(1 + 0.1)^2 = \$1{,}210$.

But suppose that we have the most customary arrangement, which is an interest and an amortization payment every year—meaning in this example the end of the first and second years. Accordingly, interest payments are still \$100 per year; but amortization payments, which will be calculated later, turn out to be \$476 per year. This is so because if \$476 is received at the end of the first year, put into a bank and transformed to $476(1 + r)$ dollars at the end of the second year, and added to the \$476 amortization payment received at the end of the second year, then the lender has $476(1 + r) + 476 = 476(1 + 0.1) + 476 =$ \$1,000 at the end of two years. Now, calling the annual amortization payment P_p, and assuming that it is unknown, it is clear that we must have $P_p(1 + r) + P_p = 1{,}000$, and with $r = 0.10$ this can be solved to give $P_p = \$476$. More generally, with L the principal, we can calculate yearly amortization payments from:

$$P_p(1 + r)^{n-1} + P_p(1 + r)^{n-2} + \quad . \; . \; . + P_p = L \qquad (7.1)$$

We can now multiply this expression by $(1 + r)$, and after subtracting it from the original expression and simplifying we get the annual amortization payment, which is:

$$P_p = \frac{rL}{(1 + r)^n - 1} \qquad (7.2)$$

If to this we add the yearly interest charge, we get for the capital cost P_c:

$$P_c = \frac{rL}{(1 + r)^n - 1} + \frac{rL}{1} = \frac{r(1 + r)^n}{(1 + r)^n - 1}\,\mathrm{L} \qquad (7.3)$$

The expression multiplying L is sometimes called the *capital recovery factor*. Furthermore, it should be recognized that the amount of money L pertains to the beginning of the initial period, and it can be regarded as the present value of a stream of payments, or the present value of some future amount F that is available at the end of n periods. This last option gives us $L = F/(1 + r)^n$, and substituting for L in equation 7.3 gives us:

$$P_c = \frac{F}{(1 + r)^n - 1} \qquad (7.4)$$

P_c can thus be regarded as one of the year-end equal payments in an n year sinking fund that permits the accumulation of a future amount F.

Next, let us understand that in a textbook world of perfect certainty, there should be no difference in the yield or return obtained through lending money (that is, buying a financial asset), and purchasing a machine or a structure (that is, buying a physical asset). However, the income from the physical asset must be such as to permit keeping this asset in such a condition that at any time it can be sold for its original price, as well as providing the equivalent of the interest income that can be realized from a financial asset. If this is impossible, at the time the physical asset is used up, or totally depreciated, it should have earned enough income to provide its owner with the same investment or consumption options that would be open for the possessor of a financial asset that was bought at the same time as the physical asset, and for the same price.

Suppose, for instance, that in this example the person lending $1,000 had purchased a machine costing $1,000 that fell apart at the end of two years. As can be easily verified, if the rate of interest is 10 percent, then a net annual revenue obtained from the machine of $576 per year will provide the owner of the machine with $1,210 at the time the machine becomes worthless. In other words, it has furnished the same alternatives for investment or consumption at the end of two years as the previously discussed financial transaction. Moreover, of this $576, $100 is designated an interest cost—it is an *opportunity cost,* and represents the interest foregone on a bank account or bond. Similarly, if we consider the remaining $476 of the two annual income payments, and the time they appear, then they accumulate to enough money at the end of two years to replace the now fully depreciated machine: $476 + 476(1 + 0.1) = $1,000. As indicated earlier, the depreciation cost ($476) is obtained using equation 7.2, while the capital cost can be determined from equation 7.3. Notice also that if this machine were indestructible—that is, had an infinite lifespan—the capital cost would be zero, and the only cost of using the asset would be the interest cost, which represents money tied up in the machine that could be used to buy a risk-free financial instrument.

Now let us calculate the capital cost for a nuclear reactor where the *investment cost* L is $420/kW, the rate of interest 9 percent, and the life of the equipment n is twenty years. We have for the capital cost:

$$P_c = \frac{rL(1 + r)^n}{(1 + r)^n - 1} = \frac{0.09 \times 420(1 + 0.09)^{20}}{(1 + 0.09)^{20} - 1} = \$46 \text{ per kW}$$

This is a useful piece of information, but we also need the capital cost of a unit of energy, and as a result time must be introduced into the analysis in a somewhat more explicit manner. If the 1 kilowatt slice of machinery mentioned above functioned twenty-four hours per day, every day of the year, then we would be talking in terms of 365 (days) × 24 (hours/day) = 8,760 hours. The $46/kilowatt would be spread over 8,760 hours, and thus the capital cost is 46/$8,760 per kWh, which is equal to 0.525 ¢/kWh, or 5.25 mills/kWh.

But we cannot expect this machinery to be perfect and function every hour of the year. Consequently the load (that is, capacity) factor—or actual number of hours that equipment is in service in relation to the total number of hours in a year—must be taken into consideration. With a load factor of 0.75, the number of hours that the equipment operates per year is reduced to 0.75 × 8,760 = 6,570. The capital cost is thus increased to 46/6,570 = 7 mills/kWh. Table 7–2 presents some average electricity costs for the OECD for the year 1977–1978, according to the International Energy Agency.

Before leaving this topic, some comment needs to be given on the rising price of nuclear electricity. The basic problem here, as alluded to earlier, is that after the incident at Three Mile Island, safety regulations became more stringent for new installations, and in some countries this meant a drastically increased construction time. For instance, in table 7.2, consider the effect of delaying the starting up of a reactor by one year. All else equal, this increases the cost of construction by at least the interest cost, or $70. Using equation 7.3 we see that the capital cost now becomes $90/kilowatt, and with a 100 percent load factor the capital cost per kilowatt-hour increases to 10.32 mills. The total cost of electricity thus becomes 2.4 + 6.5 + 10.32 = 19.22. Accordingly, even without bringing the additional construction costs into the picture, the total cost of nuclear-based electricity becomes larger than electricity generated from coal; and relatively speaking this discrepancy increases as the load factor falls.

It is also interesting to observe that the load factor for nuclear equipment in the United States in 1979 was only 0.58, which meant that

Table 7–2
OECD Energy Costs, 1977–1978

	Oil	*Coal*	*Nuclear*
Investment cost (dollars/kilowatt)	350	450	700
Operations-maintenance cost (mills/kWh)	2.0	2.2	2.4
Fuel cost (mills/kWh)	31.0	10.8	6.5
Discount rate	10%	10%	10%
Equipment life (years)	20	20	20
Capital cost (100% load factor)	4.7	6.0	9.4
Total electricity cost (100% load factor)	37.7	19.0	18.3
Total electricity cost (80% load factor)	38.9	20.5	20.6
Total electricity cost (60% load factor)	40.8	23.0	24.5

Source: International Energy Agency

Note: Up-to-date cost calculations for nuclear are given in chapter 8. Capital and electricity costs given in mills/kWh.

one kilowatt of capacity was only capable of generating $0.58 \times 8,760 = 5,080$ kWh of energy over the year. The reason for this fairly unimpressive performance can apparently be found in the large number of outages due to operational shortcomings that still characterizes nuclear facilities, in addition to the one-month period of every year that nuclear reactors must be idle so that they can be refueled. It has been claimed, however, that over time there will be an improvement in these matters, and load factors will eventually rise to 0.65–0.75.

Electricity Supply and Demand

The demand for electricity is not uniform from month to month, nor for that matter throughout a single day. In many countries there are also very dramatic shifts in the pattern of daily demand as we move through the seasons. A sample of what the daily variation might be like is shown in figure 7–1A and 7–1B. One of the most interesting aspects of these diagrams is the so-called peak, which shows the maximum or near-maximum demand for electricity. This situation is generally characterized by a high rate of capacity utilization at factories and other working places, while at the same time such things as cooking may be taking place in many households, and perhaps lights have started to come on. In Sweden, the yearly peak is in the winter, in the late afternoon, but before the end of the normal working day. By way of contrast, the yearly peak in many localities in the United States might occur on a summer weekday, late in the afternoon, but also before the end of the normal working day. This is because of the large use of air conditioners in that country. A point that can be made here is that large changes in energy prices will sometimes cause a displacement in the peak. For instance, air conditioning is to some extent a luxury, and many individuals and institutional consumers attempted to reduce its use immediately after the oil price shocks of 1973–1974 and 1979–1980.

In figure 7–1 capacity is shown the vertical axis. This simply means that we are talking about the demand for capacity, not kilowatt hours as such. Also observe that in figure 7–1B the peak—in the sense that this expression was introduced earlier—is almost obliterated for the month of July. This is unimportant, since we can define the peak as narrowly or as broadly as we wish, and as will be shown later we could specify a peak for a July day by breaking a typical day up into a peak and an off-peak period. A kind of ideal arrangement, for conceptual purposes, is shown in figure 7–1A, where the demand can be divided into three distinct parts. The baseload is the component of demand that is on line during an entire 24-hour period, although we will be forced to depart slightly

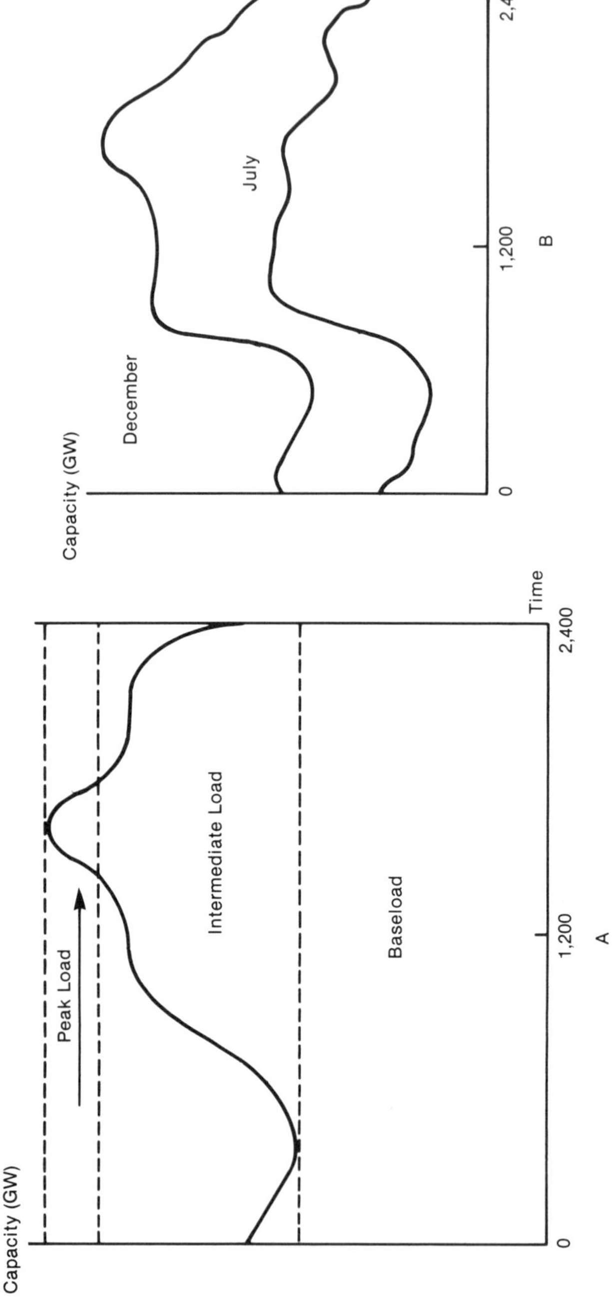

Note: 1 megawatt (MW) = 1000 kW; 1 Gigawatt (GW) = 1000 MW

Figure 7–1. A Schematic Presentation of Typical Daily Demand Patterns for Electricity in the United Kingdom, 1981

from this definition when considering the equipment that services the baseload, since this equipment must also service a portion of the load that is inactive during some of the day, and in terms of figure 7–1A is included in the intermediate load.

In the same vein, the intermediate load is a varying load that requires generation through most of the day, although, on the average, the load factor of a typical intermediate load facility seems to be between 20 and 40 percent—that is, between 20 and 40 percent of the kilowatt-hours that could be generated is actually generated. In Australia, intermediate load plants comprise about 45 percent of installed capacity, as compared to 50 percent for baseload plants and 5 percent for peaking plants. As generally designated, the peak load represents a very small segment of the demand for electricity, and usually only a fraction of the peaking capacity is in use. But even so, it should never be forgotten that the available generating equipment must be able to satisfy the maximum demand that may appear in the system, and so load factors that are palpably under 5 percent for peak load facilities are virtually unavoidable. This being the case, the capital cost of peaking equipment should be held to a minimum, since there is little or no sense in paying a great deal of money for equipment that is mostly idle. In fact economic theory tells us that, when possible, this equipment should be rented; indeed, many generating companies try to purchase their peak load requirements from similar firms in other cities, states, or countries. On the other hand, equipment servicing the baseload should have low operating costs, since functioning twenty-four hours a day can require large amounts of expensive fuel.

The way this problem is solved in practice is that various types of loads are divided up between generating euqipment having different fixed and variable cost characteristics. In Canada, the baseload requirements might be handled by a hydroelectric generator, the intermediate load by a coal power plant, and the peak load by gas turbines that can start up rapidly and accommodate fluctuating loads. The key consideration here is that while hydroelectric facilities may require a huge initial investment, operating costs are low. As table 7–3 shows, the same is true for nuclear facilities, which are also widely used for baseload generation.

In table 7–3, fossil fuel steam refers to steam raised using oil or coal as fuel. I can also mention that hydroelectric units are occasionally used for peaking purposes because of their rapid start-up time.

Now we can turn to the matter of showing how the load is divided between several types of equipment. The first thing we need is a system load (or load duration) curve, which can be obtained from daily load curves of the kind shown in figure 7–1. Formally, a load duration curve is a graphical representation of the time varying demand for electrical

Table 7–3
Capital and Variable Costs for Electricity in the United States, 1975–1976

	Investment Cost ($/kW)	*Heat Rate (Btu/kWh)*	*Fuel Cost (¢/MBtu)*	*Annual Fixed Charges per kW of Capacity*	*Operating Fuel Cost per kW of Capacity*
Nuclear steam	280	10,000	15	$42.5	1.5 mills/hour
Fossil fuel steam	210	9,000	40	$31.5	3.6 mills/hour
Gas turbine	120	13,000	70	$18.0	9.1 mills/hour

Source: *Statistical Abstracts of the United States*, 1978–1979

power. A point on the load duration curve shows (on the horizontal axis) the number of hours in a year, up to the total amount of 8,760, when the capacity to be met is at least at the level indicated on the vertical axis, which is denominated in kilowatts. Such a curve is presented in the lower frame of figure 7–2A, where the cross-over or transition hours are also indicated. The baseload equipment is in at least partial operation the full 8,760 hours of the year; the intermediate load equipment t_2', and the peaking equipment t_1'. Note that what is designated the peak load here may be part of the intermediate load in figure 7–2A. This is because the sole determinant of the transition hours is the cost characteristics of the equipment available for meeting system demand, and it may turn out that the equipment servicing the peak is also optimal for supplying a part of the intermediate load. Similarly, as mentioned earlier, it may be true that part of what had previously been designated the intermediate load is serviced by baseload equipment. In figure 7–2A, this is the area encompassed by H_1BAH_B. This situation explains why, even in theory, the load factor of baseload facilities must almost always be less than unity.

In figure 7-2A, OH_1 can be called the baseload capacity (although the part of the capacity that is used every hour of the year is only OH_B). $H_2 - H_1$ is the intermediate load capacity, while $H_p - H_1$ is the peak load capacity. Now let us examine the calculation for the transition points, t_1' and t_2'. We start by taking F_1,F_2, and F_3 as the fixed costs for gas, coal, and nuclear generating equipment, with $F_3 > F_2 > F_1$. Similarly, V_1,V_2, and V_3 are the variable costs (with $V_1 > V_2 > V_3$). The total cost of supplying a load of 1 kilowatt for t hours of the year using a plant of type i is $F_i + tV_i$, where the units are:

$$\frac{\text{Dollars}}{\text{kilowatt}} + \text{hours} \cdot \frac{\text{dollars}}{\text{kilowatt-hours}} = \frac{\text{dollars}}{\text{kilowatt}}$$

If we assume constant unit costs, then it should be obvious from figure 7–2A that regardless of the number of kilowatts we are interested in, gas turbines are the most efficient equipment for loads of the shortest duration. But eventually the time comes when:

$$F_1 + t_1'V_1 = F_2 + t_1'V_2 \tag{7.5}$$

or:

$$t_1' = \frac{F_2 - F_1}{V_1 - V_2} \tag{7.6}$$

It is also true in the situation shown in figure 7–2A that at some point we get:

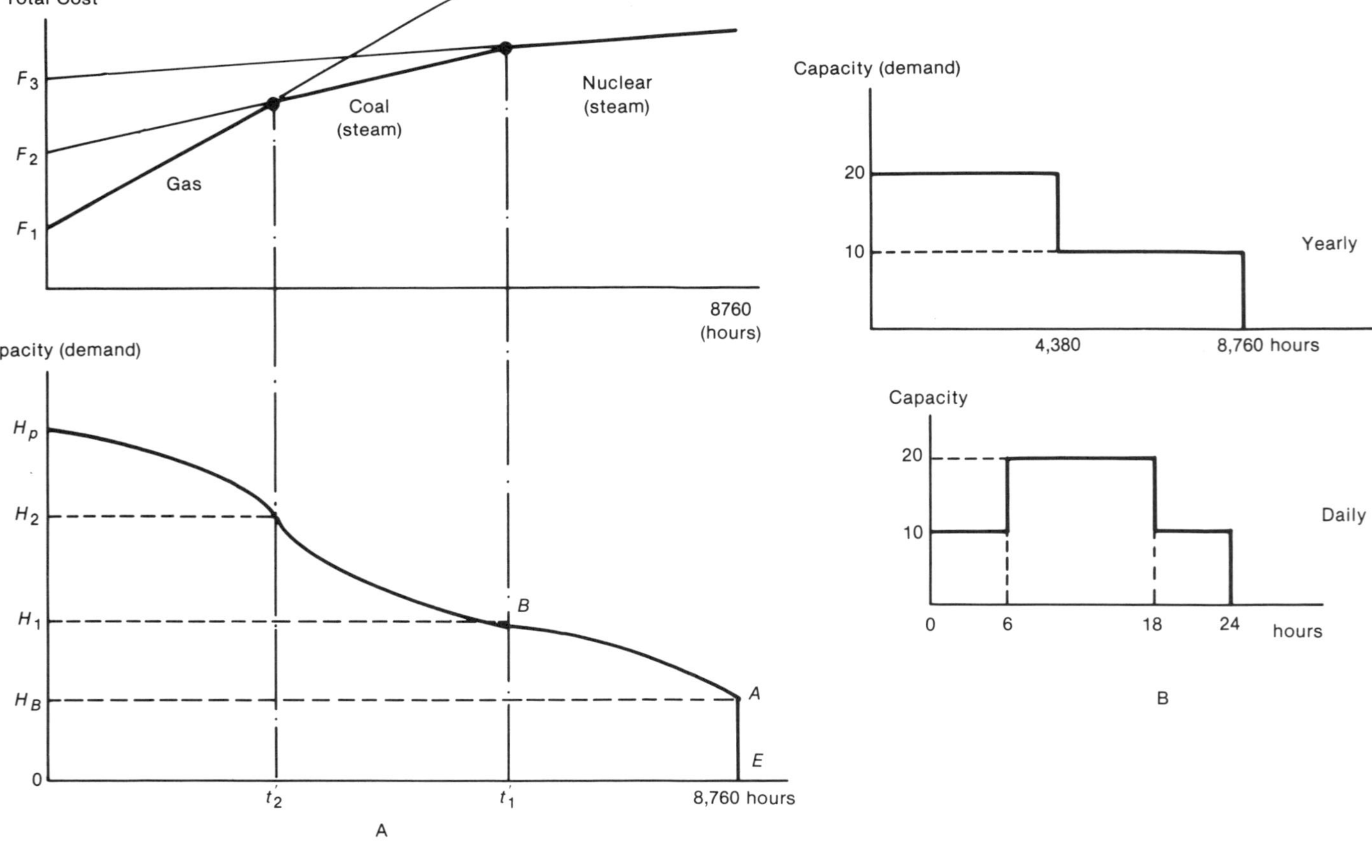

Figure 7–2. System Load Curve for Hypothetical Electricity Consumption

$$F_2 + t_2'V_2 = F_3 + t_2'V_3 \tag{7.7}$$

or:

$$t_2' = \frac{F_3 - F_2}{V_2 - V_3} \tag{7.8}$$

Something that should be appreciated here is that only cost considerations determine the cross-over points t_1' and t_2'. Demand is in the picture, however, in the form of the load duration or systems load curve. It might also be useful to point out, before presenting a simple example, that loads corresponding to the hours between t_1' and t_2' should be satisfied with a combination of base and intermediate plants. The base plant should generate as much of the required demand as possible, and therefore is operating to capacity (H_1). The remainder of the load is served by the intermediate plant.

Now let us consider figure 7–2B, where the systems load curve in the upper panel has been calculated from 365 daily load curves of the type shown in the lower panel. As can be seen, each day can be divided into a distinct peak and off-peak period; and it appears that this situation also prevails for the system load curve. But even so, as the reader can easily verify, in a situation with two types of generating equipment—for example, nuclear and gas—and the cross-over hour is to the left of 4,380 hours, nuclear plant will be used for both the base and the peak load. (If we have the total cost per kilowatt equal to $2{,}500 + t$ for nuclear, and $500 + 3t$ for gas, the cross-over time will be 1,000 hours, as can be verified from equation 7.6, for example.) Thus, even though we have a peak, we are servicing the peak with traditional baseload equipment. On the other hand, if the cross-over time is to the right of 4,380 hours, the peak will be serviced by gas while the baseload will be met by nuclear plant.

Before providing a more rigorous derivation of equations 7.6 and 7.8, it should be made clear that although the preceding analysis is extremely useful for presenting basic concepts about the optimal assortment of equipment for servicing a given load at a particular point in time, it may have very little relevance for actual planning problems. Its principal shortcomings are that it cannot take into consideration economies of scale that could result in both the capital cost per kilowatt and the running cost per kilowatt-hour varying for a given type of plant. Equally as important, it is assumed here that each generating station is capable of operating at full capacity during every hour of the year, which completely ignores the realities of forced outages and maintenance requirements (which, among other things, require the continual presence of excess capacity).

These imperfections are often reflected in the price consumers are charged for electricity. Under a time-of-day (TOD) pricing scheme, the cost of electricity depends on the margin of excess capacity: electricity used during peak periods will cost more than electricity consumed at other times. Among other things, TOD pricing is designed to induce consumers to change their temporal pattern of electricity consumption. For instance, activities such as dishwashing and the washing of clothes are, to at least a certain extent, transferable from peak hours to off-peak hours. This is more difficult where bathing and, in particular, cooking are concerned, although almost every household has the flexibility to make some changes in its scheduling of these activities.

Experience seems to indicate that the amount of load shifting possible is increased if the peak period is fairly short, while the peak price is a fairly large multiple of the off-peak and so-called shoulder rate. For instance, the scheme utilized by Boston Edison might be considered representative of this kind of thinking. From June to October the peak period is designated 11 a.m. to 5 p.m., the shoulder periods 9 a.m. to 11 a.m. and 5 p.m. to 10 p.m., while the remaining eleven hours are off-peak. The peak rate is 9.25 times the off-peak rate, while the shoulder rate is 2.58 times the off-peak rate. For the rest of the year, the peak period is from 8 a.m. to 9 p.m., with all other times considered off-peak; and the peak rate is 2.58 times the off-peak.

A Mathematical Comment on Optimal Systems Planning

By way of generalizing conditions 7.6 and 7.8, consider the total cost situation in figure 7–2A. Here we have:

$$TC = F_1(H_p - H_2) + F_2(H_2 - H_1) + F_3(H_1 - 0) + V_1\int_{H_p}^{H_2} t(H)dH + V_2\int_{H_2}^{H_1} t(H)dH + V_3\int_{H_1}^{0} t(H)dH$$

Notice that the integrations are taking place horizontally. They involve slices of the capacity H bounded by the vertical axis and the systems load curve. This type of result can be generalized to:

$$TC = \sum_{j=1}^{N} F_j(H_j - H_{j-1}) + \sum_{j=1}^{N} \left\{ V_j \int_{H_j}^{H_{j-1}} t(H)dH \right\} \tag{7.9}$$

Next we differentiate TC with respect to, for example, H_j, and set the resulting expression equal to zero:

$$\frac{\delta TC}{\delta H_j} = F_j - F_{j+1} + V_j t(H_j) - V_{j+1} t(H_j) = 0$$

From this we get as a cross-over time:

$$t(H_j) = \frac{F_{j+1} - F_j}{V_j - V_{j+1}} \tag{7.10}$$

For instance:

$$t(H_2) \equiv t_2' = \frac{F_3 - F_2}{V_2 - V_3}$$

The Price of Electricity

This section is divided into two parts. First a brief examination will be made of some of the practical aspects of energy pricing. Next, the extremely important peak load pricing problem will be reviewed as a theoretical construction—although some comments of a practical nature will also be made on this topic.

The basic issue is simple. The price of electricity is almost always an administered price. Local or national governments designate power companies as utilities (that is, natural monopolies), with the right to be the sole supplier of electricity to a region. The price that the utility can charge is usually determined or approved by a regulatory body. In general, there is an attempt to introduce prices (or rates or rate structures) that generate sufficient revenues to cover running costs, and in addition provide a competitive return on the capital used in the operation. Often this requires dividing customers into different categories (such as residential and industrial) and, as shown next, using two-part tariffs consisting of an energy charge levied on the actual kWh consumed and a capacity charge covering the supplier's readiness to provide a certain number of kilowatts.

For residential consumers in the United States, rates typically depend only on the amount of energy consumed per time period (for example, month or quarter). However, a declining block pattern is often followed, with successive blocks having a lower charge per kilowatt-hour. The intention here is to recover some fixed charges (such as those for metering) in the first, or high tariff (rate) block. By way of contrast, for major commercial and industrial consumers, there is a separate charge designed to cover the expenses associated with rather complicated and expensive metering. For example, a conventional rate structure for commercial and industrial users might call for a charge of x dollars for the first 30 kW of demand or less, a charge of y dollars per kilowatt for the next 300 kW, and a charge of z dollars for each additional kilowatt of demand. x, y, and z would be chosen in such a way as to give a declining kilowatt charge. In addition, there would be an energy charge of E_1 dollars per kilowatt-hour for the first 35,000 kWh, E_2 for the next 60,000 kWh, and a charge of E_3 for any additional kilowatt-hours of energy consumption, with $E_1 > E_2 > E_3$. It also tends to be true that industrial consumers of electricity in the United States, and many other countries, are charged for electricity at rates that are well below those charged other consumers. This is because industrial users have high load factors and generally consume electricity at higher voltages than commercial or residential consumers, which in turn means considerable cost savings on such things as step-down transformers and low-voltage distribution lines. Another point of considerable relevance for this discussion is that when price discrimination can be practiced by a monopolistic seller of electricity, those consumers having inelastic demands are typically charged more than those having elastic demands. In 1975, residential purchasers in the United States paid an average of 3.5 ¢/kWh, while industrial purchasers paid only 2.1 ¢/kWh. In the same year the elasticity of demand for industrial electricity was about -1.9, while the figure for residential electricity was -1.15. Some of this difference can be explained by the fairly wide spectrum of options available for industrial consumers, which includes the possibility of generating their own electricity. For instance, the U.S. aluminum industry consumed almost 72 billion kWh of electricity in 1979, of which about 30 percent was self-generated.

The question can now be raised as to the utility of introducing supply and demand curves into the exposition. In my opinion, there are several difficulties of a theoretical nature that make this unwise or unnecessary or both. For instance, although energy is priced in terms of kilowatt-hours, at any given moment the maximum possible load on the circuit is measured in kilowatts. Consequently, so-called supply curves for energy display a marked irregularity in their units, with kW or MW on the horizontal axis, and cents or mills per kWh on the vertical. What such a supply curve is saying is that the power company is willing to meet the demand for *x* amount of power, if the consumer is prepared to pay *y* ¢/kWh for the energy that is actually used—which is the same as pricing on the basis of possible as compared to actual demand. Since equipment must be available to satisfy any load that might be put on a line, this arrangement is understandable; but it does not correspond to the methodology featured in mainstream textbooks on microeconomics.

In the same spirit, there is the matter of defining the marginal cost at each increment of output—which is the definition of a conventional supply curve. Unfortunately, marginal cost depends on the kind of electricity generating plant in use; and although it is true that—ex ante—marginal cost can be kept low by the simple expedient of using only nuclear or hydroelectric facilities, optimal system design calls for the introduction of high marginal cost equipment into the operating cycle. However, a kind of ex post supply curve can be put together that shows the operating cost for each level of output (figure 7–3), although I personally would consider it to be of limited scientific value. Here I should add that it was the use of these constructions, and the absence of competent and systematic economic analysis, that explains the failure of the investigation of Swedish energy policy. See, for example, "Svenska statens Offentliga Utredningar: Pris på Energi (1981).

Marginal Cost Pricing

It is a fairly standard exercise in theoretical welfare economics to show that marginal cost pricing is optimal pricing from the point of view of resource allocation. When considering electricity this means that ideally each kilowatt-hour of electricity is priced at the exact marginal (that is, incremental) cost of supplying it at that moment. The question is, though, are we talking about long- or short-run marginal costs?

Unfortunately, no clear-cut answer can be given to this query. The conventional theory of competitive markets seems to indicate that price should be equal to short-run marginal costs, because these short-run marginal costs (which refer to the utilization of variable inputs) define the opportunity costs of producing more or less of the good or service we are talking about. Only in a long-run equilibrium would price also equal the long-run marginal cost. If we accept this way of looking at things, what we are going to show is that in a situation where capacity utilization is less than availability, price will be set equal to the short-run marginal cost; while when capacity is fully utilized, a price will be set at

Installation	Fuel	Capacity (MW)	Cost (mills/kWh)
Monticello	Nuclear	560	4.57
Prairie Island	Nuclear	1,040	5.92
King	Coal	574	5.92
R.B.H.[a]	Coal-oil-gas	1,215	7.16
Minnesota Valley	Coal-oil-gas	47	10.77
Wilmarth	Coal-gas	21	11.08
Lawrence	Coal-oil-gas	39	13.33
Pathfinder	Oil-gas	65	14.90
French Island	Oil	27	51.26

[a]R.B.H. signifies Riverside, Black Dog, and High Bridge.

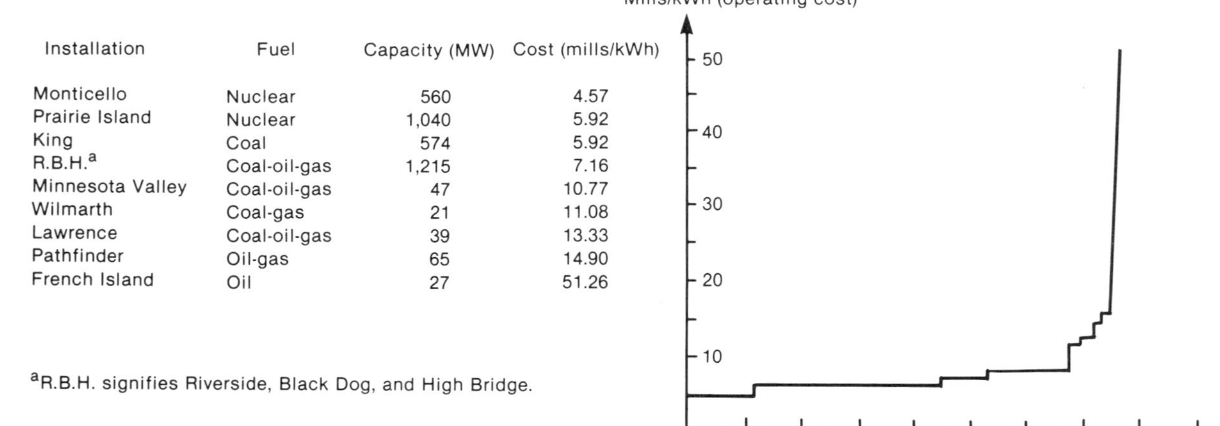

Source: U.S. Federal Power Commission

Figure 7–3. Ex Post Supply Curve for Electricity, Northern States Power Company, 1974

which the quantity demanded just equals capacity output—which implies that price can be larger than the long-run marginal cost. At the same time, the long-run marginal cost serves as a datum around which to consider additions to capacity.

It must be recognized, however, that the exact composition of the demand for electricity can be extremely difficult to estimate in advance, and in addition even the managers of electricity-generating companies (or their most technically gifted subordinates) may not be able to construct the kind of marginal cost curves for existing or planned installations that we find in reputable economic publications. Consequently, even though these managers, and the economists of regulating bodies, like to think in terms of marginal cost pricing, they often practice average cost pricing. For example, if the total annual costs associated with electricity generation (including a competitive return on capital) are \$150 million, required annual revenues will also be placed at \$150 million. If the amount of electricity that is to be delivered is 3,000 million kWh, the price of electricity will be put at 5 ¢/kWh (50 mills/kWh).

Although this may seem reasonable, there is a nontrivial economic reason for regarding average cost pricing as unsatisfactory. Consider figure 7–4 where the price p' is put equal to the average cost (at A), and so production is at q'. Observe that with production equal to q', $MC' > p'$. This is not an ideal arrangement, because in situations where resources are sold in competitive markets, the marginal cost is the social value of the resources needed to produce an extra unit of a commodity. But the value placed by consumers on the last unit of electricity sold is only p'. Thus society would be better off if the resources used to produce this unit were transferred to other uses. In fact, the same is true for all units down to q^*, and the total social gain that could be realized by reducing production to q^* is shown by the shaded area BAE.

As a caveat, it can be mentioned that the welfare loss due to average cost pricing is probably not as serious as shown in figure 7–4, since many of the resources used in the electricity-generating sector are not sold in competitive markets, and in addition are specialized to that sector. It should also be appreciated that $MR = MC$—which is the universal private profit-maximizing condition—is ignored in the present discussion, since regulatory authorities generally conclude that this arrangement yields too much profit and entails an output that is below the socially desirable.

We can now go over to some of the bread and butter issues of marginal cost pricing. To begin, consider the perfectly complementary (or Leontief) type production function shown in figure 7–5A. If we have a given (that is, existing) amount of capital $\overline{K}$, then up to the point at which this capital is fully utilized, short-run marginal cost is smaller than

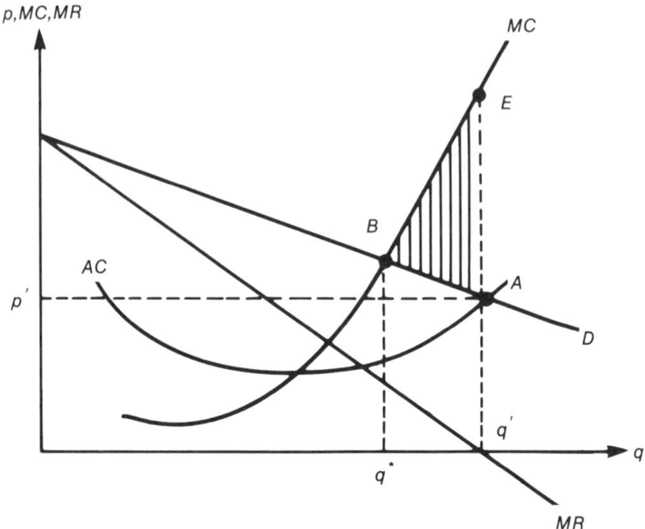

Figure 7–4. Welfare Loss Due to Average Cost Pricing

long-run marginal cost (which can also be deduced from the total cost curves in figure 7–5B). This is so because, with $\overline{K}$ given, the cost of raising production up to the point where this capital is fully utilized is only the variable cost. Once we arrive at full capacity or $\overline{q}$, to raise output we must increase both variable factors and fixed factors, which explains why the short-run marginal cost curve turns up in the manner shown. With only $\overline{K}$ of capacity, to obtain more than $\overline{q}$ of output requires an infinite increase in cost. This problem can, of course, be attacked in a neoclassical framework, but as far as I can tell, the peak load problem as such can be satisfactorily treated with the present approach, which has some obvious pedagogical advantages.

We can now observe that in figure 7–5C, several demand curves are shown; but to begin with let us assume that we have only one, D_o, which is relevant for the entire period; and on the basis of this demand curve we shall consider the price that should be charged for electricity. The period to which we are referring at this stage of the discussion is one year, and accordingly the units on the horizontal axis are kWh/year.

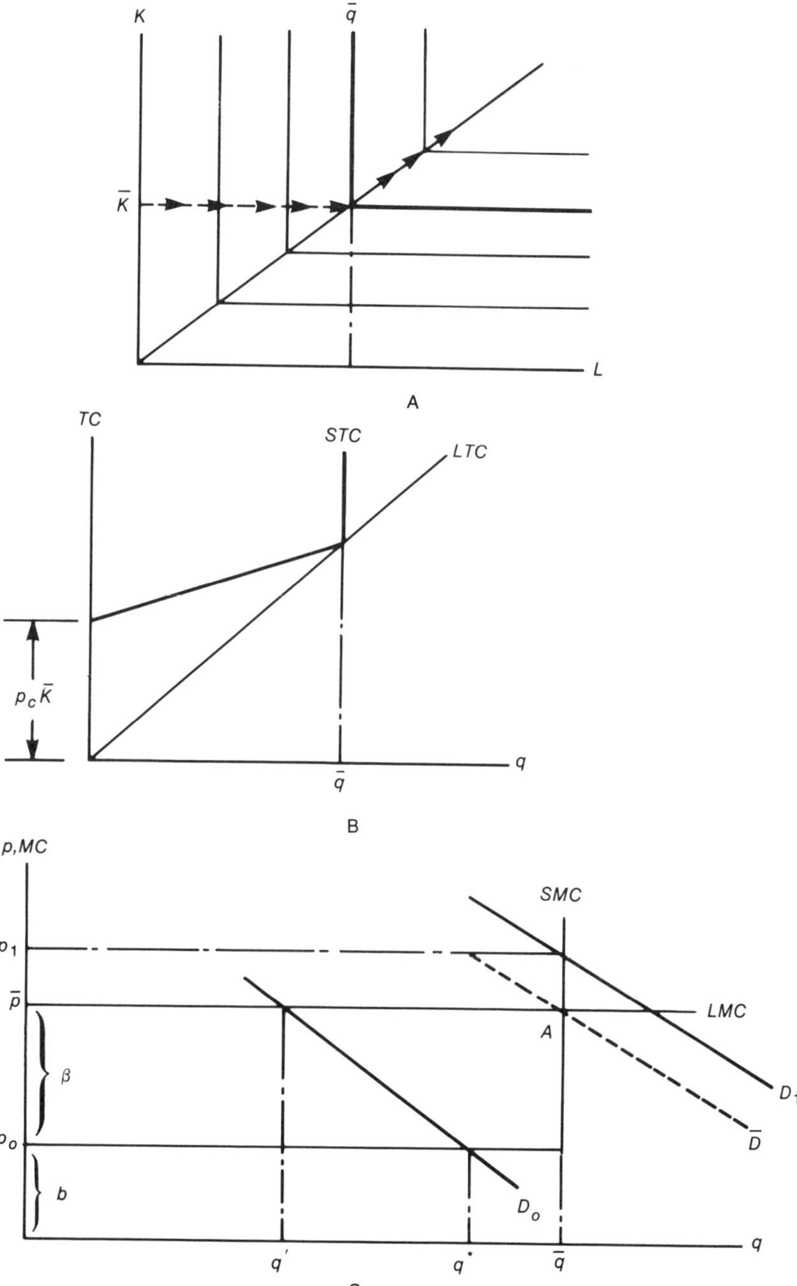

Figure 7–5. Total, Average, and Marginal Cost Curves for a Production Function Where Factors Are Perfectly Complementary

The economists of Electricité de France maintain that the correct price of electricity is $\bar{p}$, where price is equal to the long-run marginal cost (LMC). As is obvious from the diagram, output is q'. The reasoning here is that with LMC pricing, and cost curves of the type shown in the diagram, all costs are recovered. There is a large excess capacity, equal to $\bar{q} - q'$, but under normal circumstances the demand curve is moving to the right as income increases, and eventually this capacity will be taken into use. On the other hand, many economists outside France tend to view this issue in the manner suggested at the beginning of this subsection. That is, price should be equal to the short-run marginal cost (SMC). They say that since capacity exists, it should be used regardless of commercial considerations. Besides, it is generally true that the real resources used to manufacture the generating capacity cannot be recovered just because this capacity is idle, and consequently there are some important social gains that can be realized in the form of consumer surpluses by expanding production to where p = SMC (= p_o). Some of this surplus can be seen under demand curve D_o, extending downward from points $(q',\bar{p})$ to (q^*,p_o). What about average cost pricing and its no losses rationale? If, as it happens, cost curves are as shown in figure 7–5C, average cost pricing is the same as long run marginal cost pricing.

Something that can be clarified now is that the appearance of the LMC versus the SMC pricing dilemma in its more acute forms is basically due to the inability to forecast demand correctly. Were perfect foresight possible, then with a demand curve such as $\bar{D}$, capacity would be chosen so as to always operate at or very close to point A. Here capacity is optimal, with p = SMC = LMC. It is also true that if p = SMC > LMC, then additional capacity is warranted, since the value placed on an extra unit of output (p) is greater than the cost of bringing that output into existence by increasing the size of the installation. But in the short run, with SMC pricing, price = p_1.

Next we go to a state of affairs with two separate demand schedules applying to a 24-hour period. D_1 is the demand curve for the peak period (which lasts 12 hours), and D_o the demand curve for the 12-hour off-peak period. As indicated earlier, ideally rates would be different for the two periods; and with SMC pricing pricing off-peak electricity would be priced at p_o, and peak period electricity at p_1. Accordingly, the off-peak price merely covers running costs, but the peak price covers both running and capital costs for the peak period and also provides a surplus on the peak account of $p_1 - \bar{p}$. Using the logic of the previous paragraph, this particular surplus apparently indicates a willingness on the part of peak customers to pay for an increase in capacity, but unfortunately the issue is more complicated than it appears and will be taken up after a short algebraic interlude.

A Mathematical Digression on the Solution of a Peak Load Problem

At this point I would like to examine the solution to the peak load problem presented by Baumol (1977). By recasting the exercise in a slightly different framework, using duality theory instead of Kuhn-Tucker theory, it can be shown that the Baumol presentation involves a version of long-run marginal cost pricing. To keep things simple let us assume two periods, with the second period being the peak period. In each 12-hour period we have variable costs that are a function of the output and equal to $b''q_o$ and $b''q_1$ respectively. Analogously, there is a fixed cost $\beta'\bar{q}$ that is a function of the capacity output $\bar{q}$. Note that in its simplest form, β' is the 24-hour interest and depreciation cost of the capital needed to produce one unit of output. We can thus write the following linear program:

$$Z = p_o q_o + p_1 q_1 - b''q_o - b''q_1 - \beta'\bar{q} = \text{maximum}$$

or:

$$Z = (p_o - b'')q_o + (p_1 - b'')q_1 - \beta'\bar{q} = \text{maximum} \qquad (7.11)$$

with:

$$\begin{aligned} q_o \qquad\qquad &- \bar{q} \leqq 0 \\ q_1 &- \bar{q} \leqq 0 \end{aligned}$$

We know from duality theory that we have two Lagrange multipliers associated with this program. These can be defined as the shadow prices of additional capacity, and labled λ_o and λ_1. Thus we have:

$$\begin{aligned} &\lambda_o \geqq p_o - b'' \\ &\lambda_1 \geqq p_1 - b'' \\ &\lambda_o + \lambda_1 \;\; \beta' \end{aligned} \qquad (7.12)$$

This is, of course, only the dual of the original set of equations. Notice that the last expression comes from $-\lambda_o - \lambda_1 \geqq -\beta'$. Continuing, the contention that what we are dealing with here is long-run marginal cost pricing can be immediately verified if we examine the objective function of the dual. This is:

$$\lambda_o \cdot 0 + \lambda_1 \cdot 0 = \hat{Z} = 0$$

However, we know from duality theory that $\hat{Z}$ must equal Z, and so the objective function of the original (that is, primal) program must also be equal to zero. In order to bring this about, revenue must be equal to cost regardless of demand, which implies LMC pricing. Furthermore, if we have excess capacity during the off-peak period, we must have λ_o $(= \delta Z/\delta \bar{q}) = 0$. Now, from the dual program (equation 7.12), we immediately get $\beta' \geqq \lambda_1 \geqq p_1 - b''$, or $\beta' \geqq p_1 - b''$. Rearranging, we get $\beta' + b'' \geqq p_1$. In the event of production, however, $\beta' + b'' = p_1$, since $\beta' + b'' > p_1$ means that the cost of producing a unit of peak output is greater than the price of that unit, and so a linear programming solution would give $q_1 = 0$. (To avoid this rather uninteresting possibility, Baumol took the liberty of specifying a positive output in all periods.) Now, let us observe figure 7–6, which shows an off-peak production of q_o and, with the peak demand curve D_1, two possible peak period outputs, q_1 and $\bar{q}$.

In the off-peak period q_o is produced at price $p_o = b''$. In the peak period q_1 is produced, and the price charged p_1 is $b'' + \beta'$. Notice here that β' is the 24-hour capital cost, which is taken as twice the 12-hour capital cost β''. As indicated, outputs are 12-hour outputs. With productions q_o and q_1 we have an analogue of long-run marginal cost pricing,

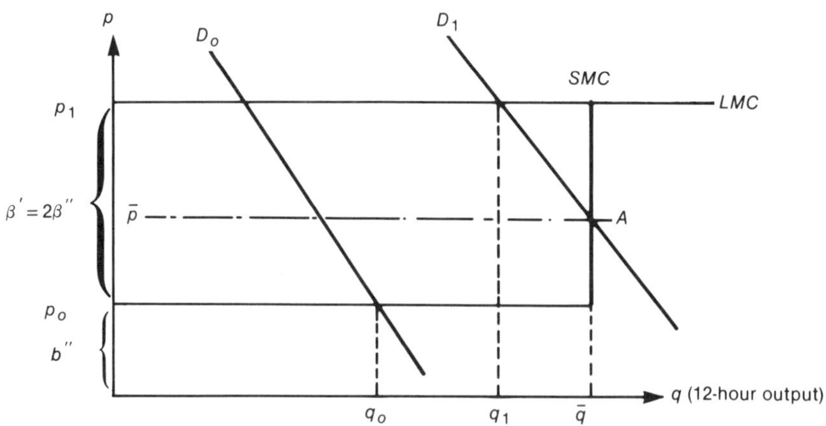

Figure 7–6. Long-Run Marginal Cost Pricing Solution to a Peak Load Problem

since all costs incurred during a 24-hour period are recovered. Had we been using short-run marginal cost pricing, the peak period output would have been $\bar{q}$. The relevant intersection of the demand and marginal cost curves would have been at A, and the price charged for peak period electricity would have been $\bar{p}$. As a result, all costs are not recovered: $p_o + \bar{p} < 2b'' + \beta'$.

Capacity Adjustment with a Differentiated Demand: An Intuitive Approach

The first item on the order of business here concerns the appropriate unit of time into which the full demand cycle should be divided. On the basis of our previous discussion it seems clear that a full day, 24 hours, defines a natural cycle, and so one option would be to divide the day into 24 periods—some peak and some off-peak. To keep things simple, the present analysis will employ two demand periods, with each being 12 hours long. Thus, the output that will be referred to is a 12-hour output. Finally, we have a demand curve for the off-peak period that will be represented by D_o, and one for the peak period that will be called D_1. The off-peak price is p_o, and the price at which electricity is sold during the peak period is p_1. In this discussion, SMC pricing will be assumed.

We will now, without further ado, write the condition that signifies the inadequacy of existing capital when demand consists of a peak and an off-peak component. This $p_o + p_1 > 2b'' + 2\beta''$, where b'' is the 12-hour variable cost of producing a unit of output, and β'' is the 12-hour capital cost. Once these two costs are completely understood, this criterion should become intuitively obvious, because all it says is that investment should take place (that is, capacity should be increased) if the value of a unit of output produced during the off-peak period, plus the value of a unit of output produced during the peak period, is at least large enough to pay the marginal costs associated with producing an additional unit of output during one or both of these periods by increasing the use of the fixed factors.

Continuing, we can take a closer look at the 12-hour costs. With the capital cost of producing one unit of output per year equal to β, and 8,760 hours in a year, the 12-hour capital cost is $12\ \beta/8{,}760 = \beta''$. Observe here that with a given $\bar{K}$ having a capacity output $\bar{q}$, then $\beta = (r + d)\bar{K}/\bar{q}$, where r is the rate of interest and d is the annual depreciation cost per unit of capital. It should also be remembered that the capital cost is something that must be paid regardless of whether the capital equipment is used to full capacity. Similarly, if the cost of increasing annual output by one unit by increasing the variable input is b, the 12-hour running cost is $12b/8{,}760 = b''$. Now let us rewrite the condition given in the previous paragraph. This gives us $(p_o - b'') + (p_1 - b'') > 2\beta''$ $(=\beta')$, where β' is the 24-hour capital cost. This says that if the excess of price over unit running costs for a unit of electricity sold during both the off-peak and peak periods exceeds the cost of acquiring the capital necessary to produce an extra unit during one or both of these periods, that extra unit should be acquired. Figures 7–7A and 7–7B can now be considered.

In figure 7–7A, off-peak production is q_o, at price $p_o = b''$; and peak period production is $\bar{q}$, at price $\bar{p}$. $\bar{p}$ $(= p_1)$ is larger than $b'' + 2\beta'' = b'' + \beta'$, and only if production in the peak period were expanded to $\hat{q}$ would this difference be eliminated. Notice also that although the acquiring of the extra capital equipment needed to produce $\hat{q} - \bar{q}$ during the peak load period means that additional equipment is available during the off-peak period, production (and demand) during the off-peak period remains constant at q_o. This result follows from the kind of production technology we have assumed, since with a perfectly complementary production function investment merely adds to the existing excess capacity during the peak period. But if we had a neoclassical production technology (with substitution possibilities), the extra equipment that becomes available during the off-peak period (because of the increase in production during the peak period) could substitute for some of the labor used to produce q_o. If the fuel used with the additional capital equipment cost less per unit of output than the other variable costs (such as labor), then q_o could be produced for a lower cost. With a fall in the price of off-peak electricity, and a demand curve such as D_o, there would be an increase in the demand (and production) of off-peak electricity.

In the situation shown in figure 7–B, the demand in either the peak or off-peak period is insufficient to cover the 24-hour capacity cost, but the sum of the off-peak and peak price is greater than $2b'' + 2\beta'' = 2b'' + \beta'$, and this is sufficient justification for an increase in capacity. Finally, the reader should pay particular attention to a certain discrepancy in these diagrams between the SMC and LMC curves. The LMC curve is, as used in the analysis, a 24-hour marginal cost curve (although in general we are thinking in terms of 12-hour outputs). Other arrangements are possible, of course, but the one used here seems easier to comprehend than the others with which I am familiar.

Real World Pricing

Unfortunately there is a difference between theory and practice, and to a certain extent this difference is unavoidable. Neither economists nor

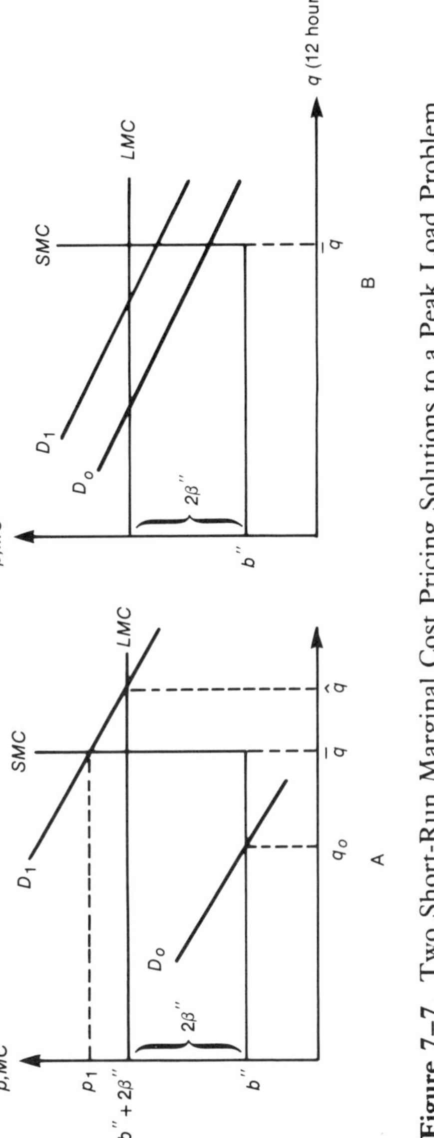

Figure 7–7. Two Short-Run Marginal Cost Pricing Solutions to a Peak Load Problem

engineers can construct usable cost curves for many actual installations; and the construction of demand curves of any type is, at least at this stage of human history, unrealizable. Furthermore, the regulatory authorities would probably not know what to do with these items if, by some miracle, they could be conjured forth.

But as it happens, people need electricity regardless of the analytical skills of economists. Since the production of electricity requires the use of real resources, it cannot be supplied free: a price must somehow be determined. Once the issue is stated in these terms, it seems clear that a practical person would insist that the price in question, whatever it is, should be capable of recovering all the costs of producing and delivering this electricity. All this sounds and is perfectly straightforward; however, I have never heard of an economist or accountant who has succeeded in devising a formula for determining the fair return to the owners of the physical capital used to produce electricity. Instead, this return is often negotiated between the regulatory authorities and the managers or owners of power companies.

In any event, once the return $\bar{r}$ has been settled on, it is applied to what is called a rate base. This rate base is the net valuation of the generating company's plant and other assets that are actually serving its customers (plus perhaps an allowance for working capital and certain other properties). The valuation of a company's tangible assets is more an art than a science, and a rough specification of the rate base is the original cost of all facilities K, adjusted for depreciation. As simple as this sounds, it conflicts somewhat with the basic tenets of capital theory, because if depreciation or other funds are being used to maintain the income-generating capacity of the asset, clearly the rate base should be related to the current evaluation of the asset. It is not impossible, of course, that this is the case in some or many localities; but the present plight of utilities in the United States suggests that some means might have to be found to expand the rate base.

We can now go to some algebra. The allowable net return can be expressed as $\overline{R}_n = \bar{r}(K - D + A)$, where A is an allowance for working capital, and D is accumulated depreciation. The expression inside the parenthesis is obviously the rate base. Next, we can define the accounting net revenue R'_n as:

$$R'_n = R_g - V - d_e - T_p - I$$

Here R_g is gross income, V signifies operating costs, d_e is depreciation calculated on a straight line basis, T_p is property taxes, and I income taxes. An allowance for deferred taxes is also usually worked into this expression, but it will be omitted here. Now we can write:

$$I = t_i (R_g - V - \bar{d}_e - E - T_p)$$

In this expression, E is interest expense, which usually denotes payments to bondholders of record, while $\bar{d}_e$ is depreciation using accelerated expensing for tax purposes. Next, setting $\overline{R}_n = R'_n$, we get:

$$\bar{r}(K - D + A) = R_g - V - d_e - T_p - t_i (R_g - V - \bar{d}_e - E - T_p)$$

This can be simplified and rewritten to give:

$$R_g = \frac{\bar{r}(K - D + A) + d_e - t_i(\bar{d}_e + E)}{1 - t_i} + T_p + V \qquad (7.13)$$

R_g can be regarded as the required gross revenue requirement in the light of the regulatory stipulations implicit in this particular analysis, and it might be instructive for the reader interested in this topic to alter the assumptions being used here. For the reader who is curious about the background of the previous derivation, it is useful to set all taxes, tax rates, and other costs equal to zero. Then we get $R_g = \bar{r}K + V$. This outcome has emerged many times in this presentation, and what it says is that the utility is permitted to recover its operating costs V plus a fair return on capital $\bar{r}K$.

Finally, a short comment about $\bar{r}$ might be appropriate. In the United States, the balance sheet of a utility can be divided into debt and equity, and an after-tax return designated for each entry. If, for example, 60 percent of the capital structure is debt, carrying an interest rate of 8 percent, and 40 percent equity, with equity owners allowed a 14 percent after-tax return on investment (which is equivalent to 28 percent pre-tax in most of the United States), then the aggregate return on investment $\bar{r}$ is a weighted average: $0.6 \times 0.08 + 0.4 \times 0.28 = 0.16\%$. Now, using this value and equation 7.3, we can get the capital recovery factor, which for a plant lasting thirty years would be 0.162 (16.2%). If the investment was \$100 million, then \$16 million would be interest and 0.2 amortization. Assuming a load factor of 0.75, the equipment would be on line $0.75 \times 8{,}760$ hours/year. If the plant is a 1,000 MW-installation (1,000,000 kW), it would supply $6{,}570 \times 1{,}000{,}000$ kW $= 6{,}570{,}000{,}000$ kWh in a year. The capital cost component of a unit of electricity can thus be calculated as $162{,}000{,}000/6{,}570{,}000{,}000 =$ \$0.02466 per kWh = 24.66 mills/kWh.

Conclusion

I have no more than scratched the surface of electricity economics; but at the same time there is at least a small amount of information in this exposition that will be of value to students of economics, economists, and others who are interested in this topic. Last, but not least, for readers desiring other elementary presentations of some of the issues considered here, I would like to recommend the work of Einhorn (1983), Jaskow and Baughman (1976), and Wenders (1976).

Appendix 7A Cogeneration

On occasion, in the United States, cogeneration has been referred to by the managers of utilities as their public enemy number one. The reason for this is easy to understand. In Freeport, Texas, Dow Chemical has completed a new power plant that generates 1,000 MW of electricity while providing the process steam that will be used in manufacturing chemicals. The local utility, Houston Lighting & Power, no longer has Dow as a customer. Instead it is purchasing electricity from Dow. What Dow is practicing is cogeneration—or harnessing the waste heat that otherwise escapes from an engine or turbine as electricity is generated. Today cogeneration produces about 7 percent of the electricity generated in the United States, as compared with 3 percent in 1981. It is believed that by the year 2000, cogeneration will be responsible for 15 percent of the power produced in that country.

A typical cogeneration facility functions as follows. Steam is raised in a high pressure boiler, and by passing this steam through a turbine it turns a shaft and generates electricity. But less than 40 percent of the energy in the fuel is transformed into electricity; the rest escapes into the air. What cogeneration does is to put the waste heat to an industrial use. When electricity prices are high relative to fossil fuel prices, considerable gains are possible from cogeneration; but at the same time, if electricity prices are low, cogeneration may not pay. Large energy-intensive industries such as paper and pulp, oil refining, chemicals, primary metals, and food processing will undoubtedly be important practitioners of cogeneration, but it is also believed that schools, hospitals, and hotels can profit from this technology.

8
Uranium and the Economics of Nuclear Energy: An Overview

This chapter surveys the economics of uranium and nuclear energy. As viewed by Hellman and Hellman (1983), nuclear energy appears to be the main competitor to steam coal, although readers who are familiar with chapters 5 and 6 of this book probably realize that I consider this a long-term view. In the short run, I feel that natural gas should play a much larger role in the energy drama, especially in Western Europe.

The precursor of today's market for uranium began with the production of the first atomic bomb during World War II, and continued after the war as a market in which sellers of uranium supplied the raw material needed to increase the world's arsenal of nuclear explosives. In North America, the demand side of the market was dominated by the U.S. Atomic Energy Commission (AEC), which also completely regulated the price of uranium. This price was immobile for long periods of time and in fact remained almost constant during the entire 1951–1962 period; but with the passage of the cold war and the rapid evolution of electricity-generating equipment employing nuclear fuels, a worldwide civilian market for uranium was eventually created. The AEC gradually declined in power, but another important agent on the U.S. atomic energy scene is the Nuclear Exchange Corporation (NUEXCO), a private brokerage firm that acts as agent for uranium transactions in the United States.

The first commercial nuclear reactor was built in Britain in 1956. It was a gas-cooled reactor installed in a so-called Magnox station. Altogether nine of these stations were put into operation by the electricity boards of Britain; but the general feeling what that this was not a successful design, nor was the second generation of gas-cooled reactors, which were called the AGR (advanced gas-cooled reactors). The nuclear bureaucracy of France also took an early interest in gas-cooled reactors, but eventually decided to adopt another type.

The most common reactor is the light water reactor (LWR), of which there are two principal variants, the pressurized water reactor (PWR) and the boiling water reactor (BWR). In a survey made in 1980 of the nuclear reactors that were operating, under construction, or on order, 289 were PWRs, 119 BWRs, and the remainder divided between about 12 different types. LWRs are a speciality of the United States and derive from its nuclear submarine program. Westinghouse is the manufacturer of the PWR, and General Electric of the BWR, and most of the LWR reactors in the world are built by these firms or under license from them. A technology that has attracted some attention lately is the Canadian CANDU (Canadian deuterium uranium) system, whose fuel is *natural* instead of *enriched* uranium. Despite pessimistic analyses of this reactor's commercial possibilities, the Canadians have been quite aggressive about building and selling it; but recently a series of accidents among CANDU reactors has put the future of the Canadian nuclear industry at risk. On the other hand, Soviet nuclear installations feature a graphite-moderated, light water–cooled reactor, and units are being built with a power rating of up to 1,500 MW.

France and the Soviet Union are the most determined nuclear advocates, and by the end of 1983 France was the second largest producer of nuclear electricity in the world. As shown in table 8–1, the United States still leads by a wide margin, but U.S. nuclear generating capacity has ceased to grow and may eventually decline. In France, licensing rests with the French government, and all French nuclear plants are produced by one supplier—Framatone—for one customer, Electricité de France. This arrangement results in a degree of standardization and speed of construction that is impossible elsewhere in the OECD, and has made a major contribution to viability in a capital-intensive activity where delays are immediately translated into large financial penalties. Customarily, reactors are built in series of almost identical units. The first big commercial reactor series consisted of 900-MW plants based on a Westinghouse PWR design. Eighteen plants were ordered between 1974 and 1980, and eventually the reactor assembly time was shortened from thirty-seven to twenty months. This meant that nuclear power plants could be built and connected to the French national grid in six years or less, as compared with ten years or more in the United States.

The Soviets have plans to increase nuclear capacity in Eastern Europe by a factor of three in the next five or six years to about 60 GW, which represents a scaling down of the original plan for 100 GW by 1990. Some people consider even the new goal impossible, although there does not seem to be any technical reason why it cannot be reached, despite a recent failure to achieve serial production at the new Atommash Reactor Factory in Volgodonsk. Interestingly enough, the French nuclear program seems to be running into some difficulties at the present time, mostly in the form of expensive plant sites, new safety features, and higher raw material costs. The cost of safety features for a new series of

Table 8–1
Nuclear Power Survey—1983 Capacity, Output, and Thermal Technology

	Nuclear Generating Capacity (GW)[1]		*Output (TWh)*		
	Installed	*Effective*	*1982*	*1983*	*Thermal Technology*
Belgium	3.5	3.5	14.8	23.0	—
Finland	2.2	2.2	15.8	16.7	—
France	27.2	25.0	103.1	137.0	PWR
West Germany	9.9	9.9	60.1	62.4	PWR, HWR, BWR
Italy	1.3	1.0	6.6	5.6	—
Netherlands	0.5	0.5	3.7	3.4	—
Spain	3.8	2.5	8.3	10.1	—
Sweden	7.4	7.4	37.3	38.9	—
Switzerland	1.9	1.9	14.3	14.8	—
United Kingdom	9.9	7.5	38.7	43.9	Magnox, AGR
Canada	7.7	7.6	40.2	50.0	HWR
United States	65.7	61.0	282.8	296.9	BWR, PWR
Japan	18.5	17.3	98.6	100.5	PWR
OECD	159.3	147.2	724.2	803.2	
Eastern Europe[2]	4.6	4.6	25.7	29.8	
USSR	19.2	18.5	102.3	106.9	
CMEA	23.8	23.0	128.0	136.7	
Other Europe	0.6	0.6	2.3	3.7	
Other Asia	6.1	5.7	18.3	29.4	
Latin America	1.6	1.0	1.9	2.6	
World total	191.4	177.5	874.5	975.4	

Source: United States Department of Energy; *International Atomic Energy Agency Bulletins: 1981–1983.*

Note: At the end of 1983, there were 282 nuclear plants in operation, and 227 on order or under construction.

[1]Installed capacity is the capacity as originally designed. Effective capacity is the average actually available during 1983.

[2]Not including the USSR.

1,300-MW units seems to have offset any economies of scale realized by these reactors, and the same is true of Electricité de France's newest 1,450-MW reactor. There is a great deal of speculation about the safety features of Soviet reactors, but two of these reactors have been sold to Finland and there has been some talk of a Soviet export program in reactors.

Next a few things will be said about the nuclear fuel cycle, and by way of introduction it should be understood that uranium has three naturally occuring isotopes: U-234, U-235, and U-238. (In this context an isotope can be taken as a component of natural uranium. The word "isotope" literally means "same place," and these isotopes occupy the same place in the periodic table of the elements, although there is a difference in their mass because they have a different number of neutrons in their nucleus.) It is U-235 that provides the atomic reactions that are the basis of the nuclear industry, because this isotope can be made to split if its nucleus is struck by a slowly moving neutron. The collision results in the disintegration of the nucleus, which in turn releases two or three more neutrons, as well as a great deal of energy. This is nuclear fission. The released neutrons then cause further fissions, and the process will accelerate. Thus we have the chain reaction that is often associated with nuclear explosions.

In general, reactors are designed to use uranium fuel which is about 2.7–3.3 percent U-235, but U-235 is only about 0.71 percent of natural uranium (that is, 1 part in 140), with the rest being mostly U-238 and a negligible amount of U-234. Thus the fuel cycle is particularly concerned with upgrading natural uranium to the required richness. Figure 8–1 shows the complete nuclear fuel cycle, and the cycle is reviewed in outline in this section. Following that we can begin the most serious work of this chapter, which is to construct a numerical example that clarifies the transformation of uranium ore into a suitable input for enrichment plants, and after that an input for a fabrication facility where the enriched uranium is transformed into a solid uranium oxide. The first four stages of the nuclear fuel cycle are called the front-end of the cycle. These are stages from the mine to the reactor core. The last two stages are the back-end, and they are designed to resolve the spent fuel problem.

The next section contains a detailed explanation of some of the operations in the nuclear fuel cycle shown in figure 8–1, but a few points should be made at the present time as a kind of introduction. Uranium is mined in either underground or open pit installations from ores having a concentration that averages about 0.25 percent if the concentration is measured in terms of uranium. (It could, for example, be measured in terms of uranium oxide, which for the purposes of this introduction can

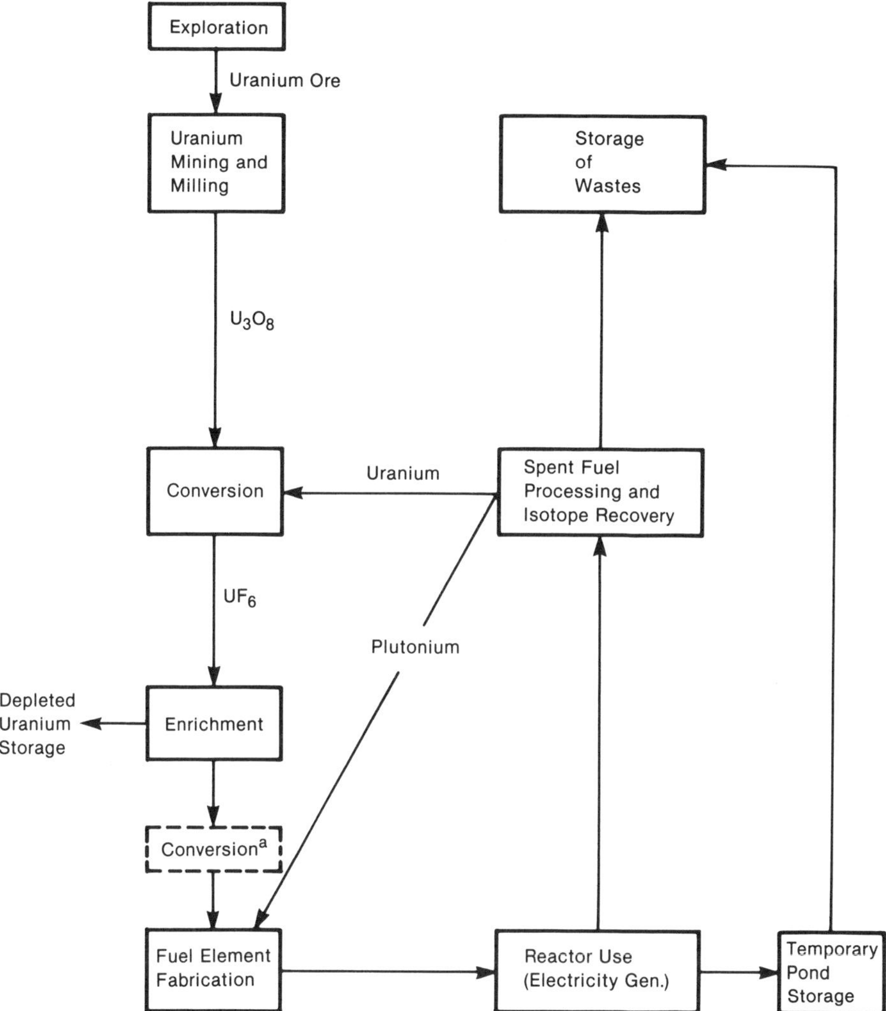

Reprinted by permission of the publisher from Ferdinand E. Banks, *Resources and Energy: An Economic Analysis* (Lexington, Mass.: Lexington Books, 1983), p. 158).

Note: This diagram does not apply to heavy water reactors (such as CANDU, SGHWR), where the moderator or coolant is heavy water. In these reactors, natural uranium is the basic fuel for the reactor core instead of enriched uranium.

*In, for example, pressurized water reactors the enriched uranium hexafluoride from the enrichment plant is converted to uranium dioxide (UO_2). After some processing these are fabricated into the fuel crates for insertion into reactor cores.

Figure 8–1. The Nuclear Fuel Cycle

be taken as a processed form of uranium.) This implies that 400 tonnes of ore must be removed to obtain one tonne of natural uranium. The general abundance of uranium is about 4 parts per million (ppm) in the earth's crust, and 3 parts per billion (ppb) in seawater. It has been estimated that about 8×10^{13} tonnes of uranium can be found in common rocks, and with known techniques 4×10^{11} tonnes of uranium can be obtained from onshore deposits. Similarly, 4×10^{9} tonnes of uranium are to be found in the oceans, and it has been said that as much as 2×10^{9} tonnes might eventually be extracted. These figures are not very meaningful, however, since uranium is only mined from deposits having a relatively high concentration.

After the extraction of the uranium ore, some elementary processing or milling transforms it into a substance called *yellowcake*, of which—on the average—from 85 to 92 percent is *uranium oxide* (U_3O_8). Yellowcake is purified by chemical means to get uranium oxide, which is converted to uranium hexafloride (UF_6), which in turn is used as a feedstock for the enrichment process that increases the proportion of U-235 in natural uranium from 0.714 percent to about 3 percent. The enriched fuel is then fabricated into the pellets or fuel rods used in nuclear reactors.

The Demand for Uranium by a Single Reactor

Now we go to one of the most important analyses of this chapter. What will be done is to specify a power rating for a nuclear plant and show the calculation for determining how much uranium must be mined to provide fuel for the plant. For example, if we have an installation with a capacity of 1,000 MW and a capacity factor ϕ of 0.65, then in a year (of 8,760 hours) we generate 5,694 GWh of energy. This is calculated as:

$$\begin{aligned} E &= C \cdot \phi \cdot 8{,}760 = 1{,}000 \times 0.65 \times 8{,}760 \\ &= 5{,}694{,}000 \text{ MWh} = 5{,}694 \text{ GWh} \end{aligned}$$

Continuing, we know that there is a certain amount of energy (measured, for example, in Btus) in a pound, gram, tonne, or any other unit of uranium, and from the last chapter we know that there is a transformation ratio between Btus and kWh. Therefore, the way seems clear for us to make the same kind of calculation we made in chapter 7. As it happens, however, nuclear specialists have simplified things by introducing the expression *burn-up*, which is a performance parameter that gives the ratio of megawatt days of electrical energy produced per tonne of enriched uranium fuel. (This concept will be explained in more detail below.) In the formula that follows the burn-up will be designated

B, and given its units we must calculate in terms of days per year (365), instead of hours per year (8,760). In addition, in our formula for obtaining the fuel requirement F, we must introduce a conversion efficiency θ, which is analogous to the heat rate employed in chapter 7. We can now write the following relation:

$$F = \frac{C \cdot \phi \cdot 365}{\theta \cdot B}$$

Examining this in terms of units we get:

$$F = \frac{\text{MW} \cdot \phi \cdot \dfrac{\text{days}}{\text{year}}}{\theta \cdot \dfrac{\text{MW days}}{\text{Metric ton}}} = \frac{\text{metric ton}}{\text{year}}$$

At this time we use the numerical data from the previous calculation. Taking $\theta = 0.333$, and $B = 33{,}000$, we get:

$$F = \frac{1000 \times 0.65 \times 365}{0.333 \times 33{,}000} = 21.786 \text{ tonnes of uranium per year}$$

This is, of course, enriched uranium (that is, containing about 3 percent U-235).

A few comments might be useful at this stage of the discussion. As mentioned, θ is analogous to the heat rate. What it says is that two-thirds of the heat content of the uranium burned in a reactor will, on the average, be lost as waste heat. Put another way, only 0.333 MW days of electricity will be generated for every 1 MW day of fuel used as an input—where the assumption is that the thermal energy in a fuel has a certain potential as electrical energy. For example, instead of describing the energy in a cord of wood in terms of Btu, it could be described in terms of kWh or MW days, and so on. The way this idea is usually expressed is that, for example, 1,000 MW of electrical energy—or 1000 MW(e)—is approximately equal to 3,000 MW (thermal). We get this by dividing 1,000 by 0.333 and rounding. Extending this discussion, we know that since 1.1 grams of *pure* U-235 contains 1 MW-day of thermal energy, in a perfect system it would be able to generate 1 MW of electrical energy; but in an average nonperfect system 1 MW-day of electrical energy requires 3 × 1.1 grams of uranium (that is, U-235) with complete fission.

Next, we can take a closer look at the concept of burn-up. If the fuel being used in this example were pure uranium U-235, fuel requirements would be 3 × 1000 × 0.65 × 365 × 1.1 = 3,000 × 0.65 × 365 × 1.1 = 782,925 grams of pure U-235, which is approximately 0.783 tonnes. (Observe that 3000 = 1000/0.333, where given capacity is 1,000 and 0.333 is the conversion efficiency). But, as just clarified, the fuel is not pure U-235; it is mostly U-238. In addition, a considerable amount of fission products are produced, and these act as neutron absorbers and reduce the reactivity of the fuel. Also fuel is customarily kept in a reactor until a significant fraction of it no longer supports the chain reaction.

Accordingly, the burn-up measures the productivity of enriched uranium (as compared to pure U-235). For example, a burn-up of 33,000 MW-days/tonne means that to produce 1 MW-day of energy in a perfect system requires 1/33,000 tonnes, or (1/33,000) × 10^6 = 30.3 grams of enriched uranium, which in terms of existing equipment corresponds to 1.1 grams of pure U-235. Using 30.3 in the above calculations (instead of 1.1) gives, as obtained previously, a figure of 21.786 for the tonnes of enriched uranium required per year to generated 5,694 GWh of energy per year. Note also in these calculations that 3 × 30.3 could represent the grams of enriched uranium required to obtain 1 MW-day in a non-perfect system (and is analogous to 3 × 1.1 grams of pure U-235).

Knowing that we need 21.786 tonnes of enriched uranium, we can turn to the details of finding out how much ore must be mined, and say something about the amount of U_3O_8 in this ore. On average, current methods of enrichment turn 1 tonne of natural uranium into 170 kilograms of enriched fuel and 830 kilograms of depleted uranium, where the enriched portion contains 3 percent (5 kilograms) of U-235, while the 830 kilograms of 'tails' (that is, residue or depleted uranium) contain 0.25 percent (2 kilograms) of U-235. Because 1 gram = 0.0022 pounds, 1 kilogram = 2.2 pounds, 170 kilograms = 170 × 2.2 = 374 pounds, and 374 pounds = 374/2205 = 0.170 tonnes. (By the same procedure, 830 kilograms = 0.830 tonnes.) Accordingly, to obtain 21.786 tonnes of enriched uranium (containing 3 percent U-235) we require 21.786/0.170 = 128.152 tonnes of natural uranium for one year's operation.

Natural uranium is obtained in the following way. An economically viable ore, usually an ore containing between 0.1 percent and 3 percent uranium, is mined. The ore is then milled (that is, crushed and chemically treated) to produce a concentrate called yellowcake, which contains uranium oxide—U_3O_8—in concentrations that average 85–92 percent. (Yellowcake can be thought of as an 'impure' uranium oxide.) The basic chemistry here says that 84.8 percent of the U_3O_8 will be natural uranium, and so if we know how much natural uranium we will need, we also know (on the average) how much U_3O_8 and yellowcake is required. At this point it might be useful to point out that the uranium mining industry formally stops with the production (in yellowcake form) of U_3O_8, and the price quotation for (natural) uranium is usually a quotation for U_3O_8. For instance, in 1979, Australian mines were selling U_3O_8 for $60 per kilogram.

Using these data, the next step is simple: 128.152 tonnes of natural uranium means 128.152/0.848 = 151.12 tonnes of U_3O_8. It also means between 151.12/0.92 = 164.25 and 151.12/0.85 = 177.7 tonnes of yellowcake. As for the amount of ore that must be mined, if we needed 128.152 tonnes of natural uranium, and ore grade (as a percent of natural uranium) was 0.1, then 128.152/0.001 = 128,152 tonnes of ore would have to be mined. Sometimes, however, ore grades are given in terms of U_3O_8. If, for example, the ore grade in terms of U_3O_8 was 0.1 percent, on the basis of the previous type of calculation, 151.12/0.001 = 151,120 tonnes of ore would have to be mined.

U_3O_8 is generally sold by uranium mines and mills directly to the utilities that operate nuclear reactors; however, it is not U_3O_8 that is shipped to enrichment plants but rather uranium hexafloride (UF_6), which is obtained in gaseous form by the further purification and conversion of U_3O_8. The reader may have been wondering about is the degree of enrichment. In the present exposition 3 percent was taken, but it could happen that the equipment being used was such that a greater (or less) degree of enrichment was required. Increasing the *degree* of enrichment with the same level of uranium input means that there would be a decrease in the output of enriched uranium. Alternatively, the tails assay—which is a measure of the U-235 left in the waste material or residue as a result of the enrichment process—could be decreased, which corresponds to squeezing more U-235 out of the natural uranium with the aid of more separative units. The complexity and expense of an enrichment plant are functions of the amount of separative work that has to be done, and as the tails assay falls, an extremely expensive process becomes rapidly more expensive. Clearly, in the event that a greater degree of enrichment is desired, there is a trade-off between raising the input of natural uranium and increasing the amount of separative work. The exact solution is determined by the relative costs.

Following enrichment, the gaseous UF_6 goes to a fabrication plant where the enriched uranium in the UF_6 is transformed into uranium oxide pellets. With no losses in this operation, we have the 21.786 tonnes of enriched uranium fuel needed for the reactor. Eventually the fuel is spent, and it is disposed of. This is a complicated matter that will not be taken up here in detail, although there are a few aspects of this process that the reader should be familiar with. After the initial loading of the reactor, about one-third of the reactor fuel core of a LWR is replaced annually. The highly radioactive spent fuel elements can be reprocessed, and at this stage unused uranium and plutonium can be recovered. (Plutonium does not occur naturally, but is created in the fission process. It is bred from U-238, which is therefore called "fertile." Like U-235, which is the only naturally occurring fissile material, plutonium—or plutonium 239 as it is often called—is fissile, and can be used as a nuclear fuel. Naturally occurring thorium 232 is also fertile because it will absorb a neutron to form the fissile uranium 233, which also does not occur naturally.) The uranium that is obtained from the reprocessing plant is richer in uranium 235 than natural uranium, and therefore is a valuable feedstock. This uranium may now be recycled back into the enrichment plant (or stored) and used again as a reactor input. On the average, reprocessed uranium can provide 19 percent of the fuel required for a nuclear reactor after the initial loading. The plutonium can also be stored or converted to oxide form and used to enrich uranium to form a mixed oxide fuel. The fissile component of this fuel is a mixture of plutonium and U-235, and it can provide for an additional 11 percent of the overall fuel need of a reactor. Thus, on the average, full processing can reduce the overall requirement of natural uranium by 30 percent over the life of a reactor.

This discussion can be concluded by saying something about the cost of electricity in the preceding example. A 1,000-MW (1,000,000-kW) installation with a capacity factor of 0.65 generates $1{,}000{,}000 \times 0.65 \times 8{,}760$ (hours) $= 5{,}694 \times 10^9$ kWh/year. It requires 128.152 tonnes of natural uranium per year for this reactor, which is equal to $128.152 \times 2{,}205 = 282{,}447$ pounds $= 128{,}385$ kilograms. The cost of natural uranium delivered to British reactors was $80/kilogram a few years ago, and so the yearly uranium cost is $128{,}385 \times 80 = \$10.27 \times 10^6$. The cost of each kWh of electricity is then $10.27 \times 10^6 / 5.694 \times 10^9 = 1.804$

Table 8–2
Uranium Production: OECD, NUEXCO Forecasts

	1985		*1986*	
	OECD	*NUEXCO*	*OECD*	*NUEXCO*
Australia	4.94	4.10	7.80	3.50
Canada	19.11	11.60	18.20	14.10
France	5.07	6.70	5.26	6.70
Namibia	5.10	5.00	5.40	5.00
Niger	13.65	—	15.60	—
South Africa	10.40	9.15	10.27	9.05
United States	29.90	8.80	32.11	8.90
Other	4.22	2.00	4.60	2.00
Total	94.34	51.15	101.19	53.05

Source: OECD (February 1982); *NUEXCO Monthly Report* (October 1983). U.S. Department of Energy, Energy Information Administration.

Note: All figures in thousands of short tons of U_3O_8.

$\times 10^{-3}$ = \$0.001804/kWh. Here we can go over to the use of mills, where 1 mill = 0.001 dollars, or 0.1 cents. Thus, in mills, the cost is approximately 1.8 mills/kWh. This would be increased to about 4 mills/kWh by the cost of enrichment and fabrication. As a result, the fuel cost of nuclear electricity is very modest when compared to oil, gas, and coal. However, when capital costs are brought into the picture, a great many things are changed.

The Demand for and Reserves of Uranium

The next major topic is the demand for uranium, and the discussion here will follow A. D. Owen (1983). Eight countries provide the bulk of the uranium supplies for the world outside the centrally planned countries (WOCA). Of these, the United States, Canada, and South Africa are well in the lead. Table 8–2 provides some uranium production forecasts by the OECD and NUEXCO. As should be immediately clear, the difference between these two sets of estimates is striking. If I were forced to choose between them, I would take those of NUEXCO, since I feel certain that the experts of the OECD have applied an unwarranted degree of sophistication to their work. Like many other people they probably do not understand that the forecasting arts are still in their infancy, where they are destined to remain. The principal thing to be kept in mind here is that with the exception of France and the countries of East Asia, the demand for nuclear equipment—and therefore uranium—is on a slightly declining trend.

As has already been indicated, the manufacturers of nuclear equipment in the United States are in serious trouble. Because of the increased

1987		*1988*		*1989*		*1990*	
OECD	*NUEXCO*	*OECD*	*NUEXCO*	*OECD*	*NUEXCO*	*OECD*	*NUEXCO*
7.80	4.20	6.76	5.60	6.11	7.00	6.11	7.00
16.77	14.87	15.99	15.35	14.95	15.35	13.65	15.35
5.26	6.70	5.26	6.70	5.26	6.70	5.26	6.70
5.40	5.00	5.40	5.00	5.40	5.00	5.40	5.00
15.60	—	15.60	—	15.60	—	15.60	—
10.14	8.90	10.14	8.90	10.01	8.90	9.88	8.90
30.55	8.90	29.90	8.85	28.99	8.15	28.34	7.45
4.91	2.00	4.84	2.00	4.95	1.80	5.10	1.80
98.38	54.35	95.84	56.20	93.22	56.70	91.29	56.00

time required to construct nuclear installations and have them certified safe, the cost of nuclear-based electricity is no longer in the bargain basement category. Utilities are also having a problem. Not only is the demand for electricity at a very low level, but after the Three Mile Island incident at Harrisburg, Pennsylvania, insurance premiums for nuclear-generating facilities have become extremely high. There have also been severe financing problems. In the first part of 1984, twenty-seven electric companies were unable to raise enough money from the bond and stock markets to complete the construction of nuclear facilities. (One of the reasons for this, of course, is that the spectacular failure of a power company in the United States resulted in losses for their bondholders that might reach a billion dollars or more.) Projections of additional U.S. generating capacity to the year 2000 range from 438 MW by the Department of Energy, to between 84 and 379 MW by the Congressional Research Service. On the other hand, the Environmental Action Foundation, which supports the Congressional Research Service, claims that no increases are needed. In its opinion, there is and will remain too much capacity, and higher plant efficiencies, combined with cogeneration, will eliminate the need to build additional capacity—which is completely wrong. (Cogeneration, as noted in appendix 7A, is the production of electricity and usable thermal energy from the same energy source, such as an industrial boiler.) Nuclear power now constitutes 10.5 percent of the generating capacity of the United States, but no utility has ordered a new reactor since 1978, and 109 reactors have been cancelled since 1972; and in February 1984, the Congressional Research Service concluded that in this century U.S. nuclear capacity was unlikely to expand beyond the plants currently under construction.

France and Japan (and West Germany) have also scaled down their nuclear programs, although more out of necessity than desire. The French nuclear program has made Electricité de France (EdF) one of the heaviest corporate debtors in the world. Still, the prevailing attitude in that country is that France must stay in the forefront of international atomic power. Today there are thirty-three nuclear plants in France with a capacity of 25 GW, while more than thirty are under construction, or on order. Forty-eight percent of French electricity is produced in nuclear plants, and this is scheduled to rise to 75 percent by 1990. According to the latest survey, French electricity tariffs are 20 percent less than those in the United Kingdom, the Netherlands, and Belgium; 30 percent below those of Germany; and 35 percent below those of the United States and Italy. Especially favored are industrial consumers, and French companies with a very high consumption are charged 30 percent less than British industrial consumers. One of the present dreams of EdF's planners is to commence a massive export of electricity, first to Germany, and perhaps later to England. Without this export the French nuclear-generating facilities will be operating at between 50 and 60 percent of capacity by 1990, as compared to the present rate of 70 percent.

The only part of the world in which there is a boom in nuclear construction is the Far East, and this is true even though the Japanese demand for nuclear equipment is slowing. The two countries in which the nuclear sector is expanding the most rapidly are South Korea and Taiwan. Taiwan obtains 38 percent of its electricity from four nuclear reactors, while South Korea has three reactors in operation and six ordered or under construction. By 1990 the intention is to obtain 50 percent of the electricity consumed in South Korea from atomic power. Some forecasts of nuclear-generating capacities for the world outside the centrally planned economies area (WOCA) is given in table 8–3.

Naturally, considerable uncertainty attaches to these figures; but although the nuclear boom is slowing, it is probably not over. In 1970, total WOCA installed nuclear capacity was only 18 GW(e), but by 1980 this figure had reached 120 GW(e), which was about 8 percent of total WOCA electricity generation in 1980. According to the former director of the International Atomic Energy Agency, Sigvard Eklund, this figure should rise to about 18 percent in 1990.

We can now consider the matter of fuel requirements for individual reactors. Once a 1,000-MW(e) nuclear power reactor is constructed, its fuel requirements range from 3,200 to 6,000 tonnes of U_3O_8 over its lifetime. For light water reactors, about 4,500 tonnes per MW(e) for a thirty-year life is the usual estimate. Table 8–4 shows the annual fuel requirements in tonnes of U_3O_8 for a 1,000-MW(e) nuclear reactor, taking into consideration the tails assay. This topic has been discussed earlier, although it was not mentioned that current techniques permit a tails assay ranging from 0.15 to 0.30 percent. According to Owen (1983), if laser enrichment were found to be a commercially viable proposition, this could be reduced to 0.05 percent. It may also be true that design improvements can reduce uranium input by 25 to 30 percent, but not—most experts feel—before the next century.

Logically, the next topic is reserves, but at this point a gap will be closed in the presentation. This deals with prices, because it is necessary to emphasize once more that prices are often quoted in terms of a certain amount of U_3O_8, and also that the trading of uranium features both spot and forward contracts. Typically, forward contracts are for a minimum of five years and usually much longer, while spot transactions are of only marginal consideration.

The real issue here is availability. A doubling, or even tripling, of uranium prices can be uncomfortable for buyers, but still bearable due to the inherently low fuel cost of electricity in nuclear installations.

Table 8–3
Installed WOCA Nuclear Power Capacity, 1990 and 2000
(GW(e))

Year Forecast Published	Source	1990			2000		
		U.S.	*Other*	*WOCA*	*U.S.*	*Other*	*WOCA*
1976	OECD	385	619	875–1004	1000	1480	2005–2480
1977	OECD	194	310	504–700	—	—	1000–1890
1979	OECD	157–192	217–268	374–460	255–395	579–812	834–1207
1981	Uranium Institute	138	214	352	—	—	—
1982	OECD	123–139	238–262	361–401	158–201	427–603	585–804
1982	NUEXCO	110	180	290	—	—	—

Source: OECD, *Uranium: Resources, Production, and Demand* (Paris: 1976, 1977, 1979, 1982); NUEXCO: *Monthly Report on the Uranium Market* (April 1982).

Table 8–4
Annual Fuel Requirements for a 1,000-MW(e) Nuclear Reactor
(tons of U_3O_8)

	Reactor Type			
	PWR		BWR	
Tails Assay	First Core	Annual Reload	First Core	Annual Reload
0.20	251	90.7	339	94.3
0.25	272	97.9	366.5	102.5
0.30	297.5	108	400	112.5

Sources: International Energy Agency, the Uranium Institute.
Note: Annual capacity factor = 0.65.

However, it would be intolerable for the operators of nuclear installations to find themselves contemplating a reduction in operations because of a lack of fuel. This, of course, has not happened yet, nor is it likely to happen in this century—or perhaps the next; but in the very long run uranium producers and purchasers may have to go over to the exploitation of fairly low grade reserves (less than 0.1 percent uranium, and perhaps much less).

Figure 8–2 and the accompanying data say something about the uranium price explosion that took place in the 1970s, particularly the relationship between the forward prices and the spot prices. Those buyers who purchased supplies in 1974 for delivery in 1979 were able to do so for prices that were one-half of the spot prices in 1979, where these spot prices were related to the cost of producing uranium in 1979. Accordingly, uranium producers were particularly anxious to renegotiate existing contracts, and their requests for renegotiation met a surprising degree of consumer acceptance. In fact, in 1979, almost all uranium buyers in the EEC agreed to renegotiate their contracts.

In the last half of the 1970s, many uranium producers became dissatisfied with the prevailing method of pricing uranium and the advantages they believed this gave uranium buyers. They insisted upon, and often obtained, contracts with escalation clauses, and without fixed prices. A contract was introduced that was similar to that used in the trading of copper, in that a known quantity of uranium was sold at a price related to the world market price at the time of delivery. But the contract also contained a price that might be termed an escalated base price, and the price paid by the buyer was the higher of these two prices. The disadvantage of this arrangement, from the point of view of the buyer, was that even if the world market price fell, the escalated base price could continue to rise since it was related to the costs of production among other things.

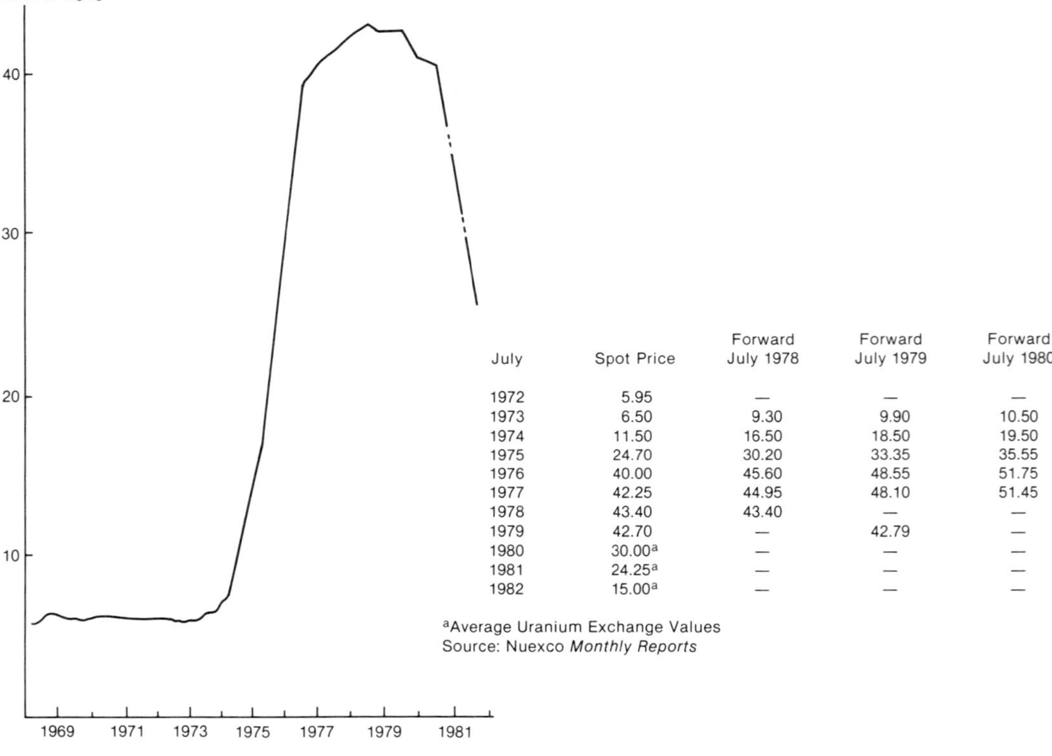

July	Spot Price	Forward July 1978	Forward July 1979	Forward July 1980
1972	5.95	—	—	—
1973	6.50	9.30	9.90	10.50
1974	11.50	16.50	18.50	19.50
1975	24.70	30.20	33.35	35.55
1976	40.00	45.60	48.55	51.75
1977	42.25	44.95	48.10	51.45
1978	43.40	43.40	—	—
1979	42.70	—	42.79	—
1980	30.00[a]	—	—	—
1981	24.25[a]	—	—	—
1982	15.00[a]	—	—	—

[a]Average Uranium Exchange Values
Source: Nuexco *Monthly Reports*

Source: Nuexco Monthly Reports
[a] Average Uranium Exchange Values
Figure 8–2. The Current Price of Uranium, 1968–1982

Finally, without prologue, we can to go the listing of reserves in table 8–5. In addition to reserves, resources are shown. In the argot of the uranium market, resources are sometimes referred to as estimated additional resources and specifically signify ores that could probably be exploited within the given cost range. Excluded from these estimations are speculative resources such as those thought to occur in undiscovered or partly defined deposits, and which are often counted as resources when considering oil or coal.

Two things need to be pointed out in association with table 8–5. First and foremost, large additional amounts of exploitable reserves are likely to be uncovered before the end of this century. In the immediate future, at least, there is plenty of uranium. In addition, the absence of the Soviet Union in table 8–5 must be explained, since that country (and the rest of Eastern Europe) has a very large nuclear construction program and it seems doubtful whether it would be willing to rely completely on the capitalist world for its uranium. It is generally believed that there is no shortage of fissionable materials within the boundaries of the Soviet Union. One problem here is that since these materials can be used for military purposes, they are not discussed openly in the Soviet Union. The Soviet Union is thought to be particularly rich in thorium, which is

Table 8–5
Uranium Reserves and Resources
(thousands of short tons of U_3O_8)

	COST RANGE ≦$30/pound ≦$66/kg		*COST RANGE $30–50/pound $66–110/R-gram*	
	Reserves	*Resources*	*Reserves*	*Resources*
Algeria	34	0	0	0
Argentina	33	7	5	13
Australia	384	30	345	27
Brazil	156	0	106	0
Canada	300	37	468	525
France	77	20	37	24
Gabon	25	3	0	13
India	42	0	1	32
Namibia	155	21	39	30
Niger	209	0	69	0
South Africa	323	142	110	119
Sweden	0	50	0	57
United States	473	317	890	543
Other	71	86	27	74
Total	2,282	713	2,097	1,457

Source: *NUEXCO Monthly Reports*, 1982–1984.

Note: Excludes speculative resources.

found in the same geological habitates as uranium, and is capable of providing a perfectly satisfactory and economical nuclear fuel.

Ownership

Some attention can now be paid to the ownership structure of the uranium industry. As to be expected, this industry is not characterized by the perfect competition arrangement of the elementary economics textbooks. Four countries account for almost 90 percent of total global production, and within these countries only a comparatively few firms are operating. Almost all of the output of South Africa is controlled by NUFCOR, which is the joint marketing company of the gold and uranium mines. The French government's Commisariat à l'energie Atomique in conjunction with the export marketing agency URANEX, virtually controls the supply of three countries—France, Gabon, and Niger—and has major ownership responsibilities in all three. Rio Tinto Zinc (RTZ) has worldwide interests in uranium mining and distribution, particularly in Namibia, Canada, and Australia. If to these organizations we add Denison Mines in Canada, and Kerr-McGee and United Nuclear in the United States, then we are talking above more than half of the production capacity in the world being under the control of six decision units.

Table 8–6 provides some additional materials on the structure of the world uranium industry, and it will be complemented by a few observations on the present capabilities of the major uranium-producing countries. We can start with Canada, whose past and committed (future) exports of uranium are the largest of any exporting country. Between 1966 (when commercial contracting began) and 1982, Canadian producers had entered into agreements to export about 120,000 tonnes of U_3O_8, of which 45,000 tonnes had already been exported. In addition, 91,000 tonnes have been committed for domestic use. Production in 1982 was approximately 9,500 tonnes, slightly less than in the previous year. The average price for all export contracts made by Canadian exporters in 1982 was $35.5 per pound (or $78 per kilogram). Spot sales amounted to only 1 percent of exports. The major producers of uranium are Denison Mines and Rio Algom (of which 52.75 percent is owned by Rio Tinto Zinc of the United Kingdom). The largest project in Canada will be that at Key Lake, which, when it comes to full production of 3,600 tonnes of U_3O_8 per year, will make it second only to the Rössing mine in Namibia (with Australia's Ranger mine in third place). The Key Lake deposit, which is owned by Saskatchewan Mining Development Corporation, Uranerz (of West Germany), and ElDorado Resources, is one of the richest large

Table 8–6
Uranium Production and Important Producers, 1979
(tonnes)

Country	*Production (1979)*	*Principal Domestic Producers*	*Main Foreign Companies*
United States	14,400	Kerr-McGee United Nuclear Anaconda General Electric	
Canada	6,600	Denison Mines Ltd. Rio Algom Eldorado	Uranerz (Federal Republic of Germany)
South Africa and Namibia	7,200	Vaal Reefs Buffelsfontein Rössing Uranium	Rio Tinto Zinc (U.K.) Rio Algom (U.K.)
Australia	500	Queensland Mines Peko Wallsend Western Mining Mary Kathleen Uranium	Getty Oil (U.S.) Western Nuclear (Canada)
France	2,300	CEA Rothschild PUK	
Niger	2,200	ONAREM	COGEMA (France) OURD (Japan)
Gabon	1,000	Cie des Mines d'Uranium de Franceville	Imetel (France)

Source: OECD, *Uranium: Resources, Production, and Demand,* the Uranium Institute, 1979.

deposits in the world. On the other hand, Rössing features an ore grade that is extremely low, but its large size and the possibility of using open cast mining has resulted in considerable economies of scale.

Canada produces almost 6 GW(e) of electrical power today in nuclear installations, and hopes to raise this to 14 GW(e) by 1991. This will mean that domestic requirements of U_3O_8 will be about 1,800 tonnes by 1992. On the other hand, Australia is strictly an exporter of uranium and does not have any domestic nuclear-generating capacity. As mentioned earlier, the Ranger deposit is now the largest producer in Australia, but the Jabiluka deposit (owned by Pan Continental and Getty Mining) was at one time scheduled for a yearly production of 4,500 tonnes, and now that the Australian government will probably change its uranium policy, it can conceivably happen that this production will be reached one day. Another large deposit is found at Olympic Dam in South Australia. This property is owned by Western Mining and BP of Australia, and plans call for production at this site of around 3,000 tonnes/year.

In 1982, twenty uranium mills were operating in the United States, of which six were working at full capacity. In 1980, the production of uranium oxide reached an all-time peak of almost 20,000 tonnes, but began to fall almost immediately, reflecting the fall in demand that accompanied the worldwide recession beginning in 1980. NUEXCO exchange values also fell sharply in real terms, and as shown in table 8–2, NUEXCO expects that many high-cost producers will have to leave this industry in the near future. U.S. consumption was 9,500 tonnes of U_3O_8 in 1982, and it has been predicted that this will rise to more than 17,000 tonnes in 1990. As a result the United States, despite its huge reserves of fissionable materials, would become an importer of uranium. Cumulative U.S. production (from 1947) reached 330,000 tonnes at the beginning of 1982, with an average ore grade over this period of 0.18 percent U_3O_8. The average ore grade at the present time is somewhat lower, but it may rise as more high-cost production is abandoned in the face of foreign competition.

Uranium prospecting, production, and use are serious matters in France, where the intention is to provide an independent source of fuel for one of the world's largest nuclear programs. Unfortunately, uranium deposits in France are small and not particularly rich; but French companies have been active in the uranium-rich francophone countries of Gabon and Niger, and they are also investing in uranium-producing capacity in Canada and Australia. In 1982 uranium production in France was 2,875 tonnes of U_3O_8, and 2,920 tonnes were imported from Gabon and Niger. France does not export uranium.

There are five major uranium deposits in Gabon (although one is effectively mined out) and two in Niger. France has substantial equity interests in the uranium mines of both these countries, but probably more so in Niger than Gabon—since Niger's domestic economic situation is not particularly inviting. However, the area to the west of the Air Mountains in Niger may be one of the richest uranium provinces in the world. Both the Japanese and Koreans are active in Gabon, while ownership interests in the uranium deposits of Niger are held by the Federal Republic of Germany, Italy, Japan, and Kuwait—in addition to France.

Finally, in the Republic of South Africa, uranium is a byproduct of gold, and of fairly minor importance. The average recovery grade is only 0.01 percent, but this is still low-cost uranium because it is a byproduct. Unlike most producers, South Africa sells much of its output on the spot market. South Africa is also very active in the Namibian mineral sector, and has control of the Rössing mine, whose production is apparently the largest in the world. This is the only mine in operation in Namibia today, but there is also a large deposit at Langer Heinrich, where the South Africans are believed to be operating a pilot plant.

Levelized Capital Costs and Load Factors

At several places in this book I have taken considerable trouble to derive and clarify the concept of capital cost. Unfortunately, the matter must be taken up again because of the need to introduce or emphasize some terminology that is common in the nuclear field, and also to serve as a prelude to a discussion on load factors. The capital cost is now termed the levelized capital cost (LC), and is defined as the cost (in dollars/kWh) that, if charged for every kWh over the life of a reactor, would pay enough in interest and amortization costs to cover the installation cost (that is, investment cost) of the reactor. For Q kilowatts, levelized cost can be defined as:

$$LC = \frac{CF \cdot K}{Q}$$

Here K is the investment cost. In the numerical example given next, calculations will be on the basis of 1 kW of capacity. CF in this formula is the capital cost factor and, as defined earlier, is:

$$CF = \frac{r(1 + r)^n}{(1 + r)^n - 1}$$

In this expression r is the discount factor and n the length of life of the installation. It should also be noted that if there are such things as decommissioning charges or property tax payments, their present value should be obtained and added to K.

An example might be useful here. Assume a nuclear plant with an investment cost of $1,800/kW, which was the average installed cost per kilowatt of nuclear-generating capacity in the United States in 1983 (in 1982 dollars). For a discount rate take $r = 10\% = 0.10$; and assume a 25-year length of life for the installation. We thus have:

$$CF = \frac{0.10\,(1 + 0.1)^{25}}{(1 + 0.1)^{25} - 1} = \frac{0.1 \times 10.834}{9.834} = 0.110$$

With an investment cost of $1,800/kW, the capital cost is $1{,}800 \times 0.110 = \$198$ per kW per year. In line with this formulae, this must be put in relation to the electricity produced per year. Taking a capacity factor (or load factor) of 0.65, the electrical energy produced by 1 kW of capacity is $1 \times 0.65 \times 8760$ (hours/year) $= 5{,}694$ kWh. Thus the capital cost component of total generating costs, or the levelized capital cost, is (198/

5,694) × 1,000 (mills/dollar) = 34.77 mills/kWh. To get the total cost per kWh, such things as the fuel, decommisioning, and insurance cost would have to be added, in addition to taxes. This would put us close to 50 mills/kWh, which is high. Several years ago the U.S. Department of Energy's Energy Information Administration (EIA) concluded that the cost of nuclear-based electricity in the United States was 43 mills/kWh (as compared to 41.3 for coal). But at that time investment costs for nuclear reactors were about $1,100/kW. By 1990 these costs may be almost $4,000/kW (against $500/kW in 1972). The investment cost for coal was also $500/kW in 1972, and it is slightly under $1,000/kW today. Here it can be noted that much of the great increase in nuclear investment costs is probably due to the increased time required to construct these installations. Even if the total labor and equipment costs for an installation taking twelve years to build were the same as for one taking six years, the total cost of the installation taking twelve years to build would still be higher because of the money tied up in the project during the period of construction. Consider, for instance, a plant whose construction began one year ago, costing $1,000—with payment to production factors today when the plant goes into operation; and a plant taking two years to build, with payments to the factors of $500 a year at the end of the first and second year of construction, and this plant also going into operation at the present time. In both cases, the investment cost appears to be $1,000, but in the second case the investment cost is actually $1,050 if the rate of interest is 10 percent. This is so because had $500 been put into a bank one year ago, instead of being used to pay for the capital and labor (and raw materials) used during the first year of construction, it would have grown to $550 at the present time. This is what we mean by money being tied up in a project.

The importance of the discount rate should also be emphasized. The more capital-intensive a project, the faster the increase in its cost as the discount rate is increased. At discount rates of 3 or 4 percent, it would be difficult to convince me that a large coal-, oil-, or gas-fired power station would be a less expensive source of electricity than a large nuclear power station anywhere in the United States. (This, incidentally, is an argument for nuclear energy, although in the conclusion to this chapter I give what I think is an important argument against.)

Now we can go to the capacity or load factor. Obviously, if nuclear plants could generate more electricity during a given year, unit costs would decrease. For example, had the load factor been 0.95 in the previous exercise instead of 0.65, the levelized capital cost of electricity would have been 33.3 mills/kWh. This kind of capital cost would be judged irresistible by most power companies.

Originally, the nuclear industry set extremely high standards for itself. These standards still exist in Sweden, and perhaps a few other places, but certaintly not everywhere in the world. Originally it was assumed that nuclear plants would generate power 80 percent of the time; now it appears that 65 percent is closer to the truth. In 1980, which was the last year for which complete data on the global stock of reactors was published, the average world load factor (specifically defined as the ratio between the energy that a power plant has produced during the period considered and the energy that it could have produced at maximum capacity under continuous operation during the whole of the period) was 62.4 percent. This represents a loss of 23,000 MW of capacity in terms of the 80 percent load factor that was once envisaged. (These figures are, of course, only for the WOCA countries.)

Between 1971 and 1980, there were about 9,000 reported breakdowns in WOCA countries, which cost more than 1 million hours of operating time, or about 550 TWh of energy. There is, however, at least one country that seems to have escaped the dilemma of a low load factor, and that is Finland. About 40 percent of the electricity generated in Finland originates with its set of four nuclear reactors. Two of these reactors (of 440 MW each) were purchased from the Soviet Union, while the other two are BWRs of Swedish design with a capacity of 660 MW each. The average load factor for the Finnish nuclear-generating sector is more than 80 percent, and the expected load factor on one of the Russian reactors and one of the Swedish reactors is almost 90 percent. The principal explanatory factor for this amazing performance, however, is not Russian workmanship or, for that matter, Swedish engineering abilities, but the traditional feeling of the Finns for such things as quality. This means that, in theory, higher load factors are possible everywhere, because in every country it is possible to find people who are willing to set high professional standards, and live up to them. If this were not true, the airlines would not be able to function.

Before concluding this section, two expressions should be defined. One is *operating factor,* which is the ratio between the number of hours a power plant is on line to the total number of hours in the reference period. The other is *unavailability factor,* which is the amount by which the available capacity is lower than the maximum capacity. There are two unavailability factors, with one applying to planned outages for refueling and maintenance; and the other with unplanned outages (breakdowns).

Uranium Inventories

The final topic in this chapter concerns uranium inventories. The problem here is approximately the same for all fuels and minerals.

Inventories are held for both speculative and transactions motives. They are held because it is expensive not to meet contractual obligations, and so if something goes wrong in the uranium delivery chain, a buffer must be available; similarly, holding stocks is a good hedge against expected price increases, and occasionally these stocks can be sold for a very good profit. The key concept where this subject is concerned is the equality, or lack of equality, of planned and actual inventories. When planned inventories are greater than actual inventories, an attempt will be made to increase inventories, which will often put an upward pressure on the price of the commodity.

In considering the uranium market, however, the opposite situation prevails at the present time. According to NUEXCO, uranium production will exceed consumption throughout the 1980s, although the extent of the excess supply will probably decline. Moreover, there will probably be a relative fall in inventories in the United States during the rest of this decade, since a DOE survey found that a majority of U.S. utilities with specific policies regarding inventories consider two years of forward coverage for U_3O_8 to be satisfactory, although on the average they are holding five years coverage in stocks. Table 8–7 gives some NUEXCO data for production, consumption, and inventories of U_3O_8, and projections for 1985 and 1990.

As I emphasized in *The Political Economy of Oil*, inventories are central in the matter of price determination—so much so, in fact, that the simple flow modes of the microeconomic textbooks can, at best, provide only a partial explanation of price movements in this type of market, although many writers of this type of textbook do not seem to realize this fact. Also, uncertainty is always present in the kind of model to be described now, because without uncertainty inventory holders would never find themselves with too much or too little inventory.

At this point I submit figure 8–3 as an alternative to the simple flow

Table 8–7
Uranium Production, Consumption, and Inventories, 1982–1990
(thousand of short tons)

	1982	*1983*	*1984*	*1985*	*1990*
Production (s)	53.8	52.5	53.0	51.1	56.0
Consumption (d)	28.3	37.6	38.3	40.6	48.2
Net stock increase	25.5	14.9	14.7	10.6	7.8
Total stocks (I)	238.6	253.4	268.1	278.7	315.0
I/d	5.5	5.6	5.7	5.7	6.3

Source: NUEXCO *Report* (1981–1983).
Note: S represents flow supply, d flow demand, I total consumer and producer stocks.

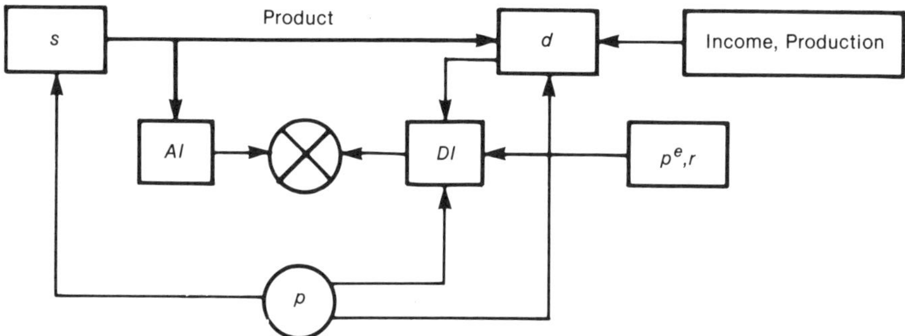

Figure 8–3. Supply-Demand System Showing the Interaction of Desired and Actual Inventories

model of the elementary textbooks. The thick line shows the movement of product from supply s to consumption d. ('Demand' is defined as the demand for current consumption *and* the demand for inventories). Actual inventories are represented by AI, while desired inventories are DI. Desired inventories are taken as a function of such things as the current price (p_t), the expected price (p_t^e), the interest rate (r), and so on. These are also indicated in the figure. The rules determining price formation are as follows: If $DI = AI$, then price is stationary, or $p_t = p_{t+1}$. But if $DI > AI$, then $p_{t+1} > p_t$, since the price will have to be raised in the next period to get producers to produce more of the commodity or consumers to lower their current consumption. Similarly, if $DI < AI$, price will fall as inventory owners put their excess inventory on the market and so $p_{t+1} < p_t$. Clearly, it is possible to postulate a relationship of the form $p_{t+1} = p_t + \lambda(DI_t - AI_t)$, with $\lambda > 0$.

The feedback in this model should be carefully studied. Price changes come about because of changes in the demand for stocks, but these price changes also influence current production and consumption, unless demand or supply curves are inelastic. There is accordingly an interaction or interplay between stocks and flows (where an example of a flow would be the amount being currently produced or consumed, and whose dimensions are quantity per time period). This type of model is therefore a stock-flow model, while the models used in most microeconomics textbooks are flow models. Also, in stock-flow models it is necessary to distinguish between two types of equilibria: *temporary* or *market* equilibria, in which the desired additions to or subtraction from existing inventories in a given period is equal to the actual change; and *full equilibria,* when current supply is equal to current demand, and actual inventories are equal to desired inventories.

Finally, a comment on today's uranium market. At present there is an abundance of uranium on world markets, and if it is true that there will be an increase in the supply of low-cost Australian uranium in the near future, a number of high-cost producers in the United States will not be in business much longer. It is expected, however, that utilities in the United States possessing what they consider to be excessive stocks of uranium will be rid of these around 1987 or 1988. As a result the uranium price should eventually be able to climb back to somewhere in the \$35–40 per pound (of U_3O_8) range. Until then it will languish below \$25 per pound.

Conclusion: A Personal Opinion

I have attempted in this chapter to provide a brief, but complete, survey of the elementary economics of uranium and nuclear energy. What I have not done, thus far, is to offer a personal opinion on the overall suitability of nuclear energy, other than to say that (excluding hydropower), it is the most inexpensive source of electricity almost everywhere, and it could be the most inexpensive everywhere, including the United States.

Apart from that, however, I distrust uranium. I do not believe that most parliamentarians or nuclear bureaucrats have the intellectual capacity, imagination, patience, or inclination to study and fully comprehend the intricacies and ramifications of uranium (and radioactivity). These ladies and gentlemen are, in general, incapable of devising rules for the safe use of uranium, and at the same time they are incapable of understanding what a failure to devise such rules will mean for the future of our civilization. Certainly this is true in Sweden, and it is probably true everywhere. There is also the rather delicate matter of technical competance.

Swedish nuclear reactor manufacturing and technology, as well as the operation of nuclear installations, undoubtedly has met or exceeded the highest technical and administrative standards in the world. At the same time, considering the popularity of the mediocre scholars, bureaucrats, and celebrities who have or will become policymakers in the Swedish version of Academia, it is only a matter of time before this kind of blunderer will be in positions of responsibility in the nuclear sector. In fact, to a certain extent, this situation already exists, because the Swedish energy bureaucracy is, and has always been, everything that the U.S. Department of Energy was accused of being (but was not)—only much, much, much worse.

Certainly, some chances must be taken with nuclear energy, given the unemployment figures that prevail throughout the industrial world. But, thank you, as few as possible. And none involving plutonium.

9
Summary: The World Coal Market

This book is intended as a rudimentary but thorough expedition through some of the intricacies of applied energy economics. The main topic has been coal, but I especially hope that readers have found the surveys of oil, natural gas, nuclear energy (and uranium), and electricity helpful. This concluding chapter will present a short up-to-date survey of present trends in the world coal market.

In the panic that followed the first oil price shock, it was predicted that a general *hausse* was in store for the world coal market. According to the celebrated World Coal Study (WOCOL) of 1976, Western Europe would be using 416 million tons of coal (equivalent) in 1990, an estimate that now appears to be at least 20 percent higher than the actual consumption that will be reached at that time. Similarly, in 1982, the International Energy Agency (whose members are the OECD countries minus France, Finland, and Iceland) predicted an 85 percent rise in consumption for its member countries in the period 1980–2000. This estimate has now been reduced to 60 percent, which some observers regard as completely beyond reach. In fact, on at least one occasion, the *Financial Times Energy Economist* (February 1984) claimed that the IEA has been attempting to talk up a market for coal that, in reality, does not and will not exist in the foreseeable future.

The main problem, as pointed out in chapters 2 and 3, is that many potential consumers of coal will do everything possible to avoid using it. If, for example, there is going to be an ample supply of oil available at reasonable prices—and this is the message being flashed to many potential coal users by the present excess supply of oil—then why return to the age of coal? Not only is coal troublesome to handle and use, but—as pointed out in chapter 6—coal buyers are aware that environmental legislation in most countries cannot be regarded as predictable. This is particularly true in the light of new information about the effects of fossil fuels on the German forests, where the phenomenon known as *Waldsterben* (forest death) has reached astonishing proportions in the past few years.

Industrial users of coal, who consume about 25 percent of the global annual coal production, have generally refused to switch from oil to coal until they have a clearer picture of the future availability of coal. Even more provocative, many of the electricity suppliers, on whom coal sellers counted so much, feel the same way. In Holland, Italy, and Belgium, a considerable amount of oil is being replaced by gas, and not coal, in power stations. This has been termed an interim measure in these countries, but thanks to the huge increase in the supplies of natural gas that resulted from the completion of the Soviet–Western European pipeline (which also depressed the price of Algerian gas), the present energy picture in Western Europe is much brighter than it was a few years ago, and this has reflected on the demand for coal. The upshot of this situation is that coal production in the United States, Canada, and Australia—the major coal-producing countries of the OECD—has fallen to twenty percent below capacity.

But the coal buyers' refusal to speed up their conversion to coal is not the complete explanation for the world coal industry's present doldrums. There is also the matter of lagging economic activity, and, in particular, an anemic demand for steel and electricity. Taking the first of these, it is interesting to note that when I was associated with the ad hoc Steel Committee of the United Nations Industrial Development Organization (UNIDO) a decade ago, the prediction was for a shortage of metallurgical (or coking) coal in the medium run, or maybe sooner, and everything possible was being done in some countries to speed research on techniques for blending metallurgical and thermal (i.e., steam) coal in such a way that the mixture could be used to replace coking coal in existing blast furnaces. At this time there is a surplus of metallurgical coal that is expected to persist indefinitely, even if new steel mills are constructed in countries like South Korea, Brazil, and Turkey that import coking coal. The problem, which I hope students of economics will carefully notice, is that on the basis of forecasts of market demand, an enormous amount was invested in facilities for producing coking coal, and this investment was encouraged and financed by the world's banking community. For example, altogether, the capital market funded two-thirds of the investment in the Gregg River, Quintette, Bullmoose, Line Creek, and Greenhills mines in Canada—leaving only one-third to be supplied by partners in these ventures. The same thing happened in Australia, where a large number of banks liberally supplied loans to both low- and high-cost mines in Queensland, believing that the market for coking coal was about to open up.

As we now know, this did not happen. Instead, beginning in 1983, coking coal sellers throughout the world were informed by their clients that they would be unable to accept an appreciable part of the metallurgical coal they had bought on long-term contracts, and eventually many suppliers had to accept price cuts of up to 20 percent on previously contracted sales. Later, further price and quantity cuts were introduced for supplies scheduled to be delivered during 1984 and 1985, although some coal-producing firms were able—at the expense of other firms—to negotiate increases in quantity in return for accepting decreases in price. What should be noted is that, except in special cases, this kind of market behavior cannot be described by any formal model in the arsenal of elementary or advanced microeconomics. Given this regrettable situation, let me assure those who have convinced their sponsors, and perhaps themselves, that the markets for coal can be modeled using a blend of game and oligopoly theory, that they are going to be gravely disappointed. It is also interesting to observe that, while many producers of coking coal are not presently making a profit, most of these firms continue to ship coal in order to retain their claim to a share in a market that they feel will eventually expand. Some companies cannot sustain this behavior, of course, and eventually reduce or terminate their commitments to their clients. But once again, in direct contradiction of the results predicted in microeconomics texts, many of the firms remaining in these markets are not necessarily those having the lowest costs, but enterprises that are subsidiaries of very large companies that happen to possess the financial resources required to subsidize long periods of unprofitable activity in anticipation of an up-swing in demand.

It should be emphasized that only a few countries have complete data on proven reserves of coking coal. The total reserves of metallurgical coal were recently estimated to be about 75 billion tons—approximately 18 percent of the world's proven reserves of anthracite and bituminous coals. About 20 percent of the proven reserves of U.S. coal is metallurgical coal, while for Australia and West Germany, metallurgical coal reserves are approximately 60 percent of total bituminous coal reserves. South African coal is mostly of the thermal variety. Since about 1980 the export price of low-volatile metallurgical coal has been $10–$20 higher than steam coal, with the exact spread depending on the state of the markets for iron and steel. Historically, metallurgical coal has tended to trade at a premium over thermal coal, and the World Bank expects this situation to continue, even though the growth-of-demand rate for metallurgical coal will probably be only 50 percent of that for steam coal in the immediate future. About 27 percent of the coal produced in the World Outside the Centrally Planned Areas (WOCA) is metallurgical coal, although 75 percent of world coal exports consists of metallurgical coal. In fact, international trade in thermal coal became economical only after the oil price shock of 1973–74.

An Overview of Thermal (Steam) Coal

This section will examine the situation of some countries that are purchasing thermal coal, but first some general comments on coal are in order.

In terms of the degree of carbonization and heat values, there are three broad categories of coal: anthracite (which is 5 percent of production), bituminous and sub-bituminous (70 percent), and lignite (25 percent). Two-thirds of world coal production takes place in the United States, the Soviet Union, and China, with most of the rest in Australia, South Africa, Poland, the United Kingdom, the Federal Republic of Germany, and Canada. India is also a large producer, and the other three Third World countries considered to have impressive long-run prospects are Botswana, Columbia, and Indonesia. The two basic methods of coal mining are underground and surface mining. The production costs for surface mining are usually much lower than those for underground mining, which is mostly due to the economics of scale that result from greater mechanization. IN 1982 the international average cost of mining a ton of coal in the so-called low-cost countries was $25–$35 a ton (in 1981 constant U.S. dollars), while it was as much as $65 per ton in a high-cost country like the United Kingdom, and between $45 and $50 per ton in West Germany and South Korea. By the same token, in low-cost countries, surface mining costs averaged $10–$20 per ton. In underground mining, the cost of labor is, on average, 50 percent of the total mining costs; capital costs are about 22 percent; while the cost of materials, such as fuel, takes up the remaining 28 percent. In surface mining, capital costs are 50 percent, while labor costs are 20 percent.

In the United States during the 1970s, the wages of coal miners increased faster than the manufacturing wage by about 2 percent per year, despite an average annual decline in productivity of 6 percent per year that may have continued to 1982. This wage pattern in the United States during these years generally resembled that of most coal mining countries, although labor productivity in coal mining outside the United States did not fall, and may have had a small average increase. At the present time in the United States, it is expected that real wages in the coal mining sector will not rise as rapidly in the near future as during the past decade-and-a-half. On the other hand, according to the World Bank, the real cost of the rail transportation of coal in the United States increased by 4–6 percent annually in the 1970s. This was largely due to the near-monopoly of the railroads in that country, and together with the lack of ports that could service the largest coal-handling vessels (called colliers), this situation has greatly reduced the competitiveness of U.S. coal in foreign markets. U.S. activity in the world market for steam coal will be taken up below, but I can mention here that I expect these disadvantages to be removed before the end of the 1980s, or shortly after.

Prior to the first oil price shock, the main reason for the decline of coal and the rise of oil was the change in the relative price of these commodities. Between 1960 and 1973 the real price of coal declined by 9.3 percent, while that of heavy fuel oil (the principal competitor of coal in heating and electricity generation) declined by 29 percent. The first oil price shock altered this relationship, but not by enough to give coal an overwhelming advantage—particularly when the environmental costs that may be associated with coal are taken into consideration. The share of coal in total primary energy consumption in industrial countries will definitely exceed the 20 percent registered in 1980, for the simple reason that it is inconceivable that oil will regain its former position. However, there are some experts who claim that unless large-scale coal gasification becomes feasible, the increase in the rate of growth of coal consumption will become very small.

In the next decade, more than three-quarters of the projected increase in steam coal demand will come from the electricity generating sector, which now accounts for 70 percent of WOCA's steam coal consumption. Most of the rest will accrue to the industrial sector. It is expected—though not certain—that there will be a gradual replacement and/or modification of the oil-fired boilers presently used for power generation, since contemporary practice generally opposes the employment of oil-fired power plants in new base load facilities. But converting existing oil-fired boilers into coal-fired installations is often judged to be too costly, particularly since oil prices appear to have stabilized. However, some research is now being carried out to explore the use of oil-coal mixes in existing boilers, which would obviate the preliminary scrapping of oil-using equipment. (See chapter 7 for a discussion of 'base load'.)

There are many coal-producing firms in the world, and for this reason many investigators have—mistakenly—attempted to treat the international coal market as if it approximated the perfect competition paradigm of the microeconomic textbooks. My contention, however, is that this approach to the world coal market is deluded. The law of one price does *not* always apply in this market, and there are plenty of cases in which buyers have purchased at high cost instead of at low cost coal because they thought this would ensure future access to large quantities of the commodity. In the United States, though, where there are more than 3,000 mines contributing to the aggregate supply of coal, it is possible that some local markets in or near the coal mining districts may display many characteristics of the ideal perfectly competitive markets. (At the same time, the reader should be aware that in 1977 in the United States only fifteen companies accounted for 41 percent of the country's total production of bituminous coal, while the top fifty companies produced 65 percent of the total. This disposition no longer prevails, however, since the upswing in coal demand during the 1970s led to the expansion of many small firms. Chapter 4 provides some further insights into this matter).

The eastern and midwestern mining regions of the United States are distinguished by the presence of hundreds of small and medium-sized coal producing companies. These regions provide most of the U.S. exports of steam coal to Western Europe, because high rail tariffs have tended to reduce the access of western coal to eastern and Gulf coast ports. (South African steam coal shipped by sea to the U.S. Gulf coast has been less expensive than western coal shipped by rail, and the development of new coal fields in the western states could entail some rail hauls that are approximately twice as long as current hauls). Another important but often overlooked aspect of this topic is the expected change in the cost of low-sulfur eastern coal. Assuming constant factor costs and no technological change in mining, the long-run marginal cost of Eastern low-sulfur coal—the portion of eastern coal entering into international trade—might increase rapidly as production rises. On several occasions in this book I have pointed out that if the United States is to fulfill its promise to be the Saudia Arabia of thermal coal, then the output of low-sulfur, low-cost western coal must expand considerably. If not, then potential coal users, particularly in Europe, might conclude that the United States will be unable to supply their requirements in the long run at prices that are competitive with other energy materials.

European Importers of Coal

This section examines the behavior of the main importers of coal, concentrating especially on France, which until its nuclear program recently accelerated was Europe's leading coal buyer. France is also important not only because it is in the process of becoming the world's leading consumer of nuclear-based electricity, but because it is a paragon of energy conservation. France's basic intention is to reduce dependence on outside fuels to a minimum, which means eliminating as much coal and oil as possible in the production of electricity, and at the same time making sure that all fuels are burned more efficiently. By May 1984, the energy conservation program formulated in the mid-1970s had resulted in France saving more energy per capita than any country except Denmark. According to a report of the European Economic Commission (EEC), Denmark had reduced its 1973 energy consumption by 33 percent, and France by 25 percent.

A new program, formulated in 1981, calls for even more radical changes: by 1990, energy consumption is to be cut by the equivalent of 40 million tonnes of oil, with half the conservation taking place in the housing and commercial sectors, while industry and transport each try to save 25 percent. The goal is for oil to represent only 30–32 percent of the energy used in 1990, as compared to 48 percent in 1981, and 66 percent in 1973. Renewable sources of energy (such as solar, biomass, wind, and thermal) are to triple their 1982 contributions, reaching the equivalent of 10–14 million tons of oil in 1990. On the macroeconomic front, the investments required to attain this goal are about 50 billion francs per year; but by 1990 the negative entries on the balance of payments (which to a considerable extent have to do with energy) are expected to be decreased by about 440 million francs per year. In theory, with the gradual removal of energy's constraint on economic growth, 330,000 jobs can be created by 1990, while gross domestic product will be able to increase by 3–4 percent per year.

I certainly hope that these expectations can be fulfilled, because up to now a number of countries have made the mistake of believing that energy can be replaced by something other than energy. In Denmark, for instance, the theory was that a low-energy community enjoyed the same economic prospects as one employing much more energy per capita, and as a result unemployment spiraled almost completely out of control, while the standard of living fell for many employed people. It has yet to be proved to my satisfaction that Denmark—or any other country in Western Europe—will be able to regain pre-1973 rates of economic growth (and progress) on a low-energy diet. The French, however, are intent upon reducing the inefficiencies in energy consumption as much as possible, while making sure that aggregate energy intensity is not reduced to a point where the standard of living is threatened. Needless to say, a policy of this kind not only requires a great deal of imagination and skill on the part of policymakers and their economic advisors, but also an almost mystical faith in the ultimate optimality of nuclear power.

The official coal importer, ATIC, estimates that French coal imports in 1984 will only reach 18 million tonnes, as compared to 20.2 million tonnes in 1983, 24.8 in 1982, and 30.2 million tonnes in 1981. Domestic coal production at present is 18.5 million tonnes/year, although an election promise of President Mitterrand was production of coal at 30 million tonnes/year for 1990, and therafter. As things now stand, some observers are predicting a decline in domestic production to 15 million tonnes/year, and perhaps lower. The largest cuts in domestic purchases have been made by Electricité de France and the steel industry, while the Republic of South Africa appears to be the country whose sales to French coal importers has fallen the most. It also seems that Poland has increased sales to France over the past few years. (See chapter 3 for information on the Polish coal industry.) If decreasing the size of the coal sector were not a sensitive political problem, French collieries would now be closing down at a record pace, since the government's intention, and the energy bureaucracy's, is to use as much nuclear energy as possible. In addition, many government planners and decision-makers have a basic distrust of the United States' ability to provide Western Europe with the quantity of coal that the International Energy Agency has claimed that the United States is capable of supplying. Quite simply, the French want to know whether the United States can deal with the massive environmental disturbances implied by raising their exports to the level suggested by the IEA. Concomitantly, once coal becomes the key factor in Western Europe's energy picture, is it certain that nothing will happen to interrupt deliveries if these environmental disturbances become too severe? This question cannot, unfortunately, be answered in this book, although in my opinion the United States will—in the long, though perhaps not the short run—prove capable of supplying enough coal so that, *ceteris paribus*, the threat of further energy price shocks will be removed.

As is to be expected, some people view the issues discussed above in quite another fashion. Christian Gerondeau, in his book *L'Energie à Revendre* (Paris:Lattès, 1984), concludes that the energy crisis is over and that from now on there will be a surplus of energy. The only comment I can make here is that Gerondeau is an engineer, and not an economist; and while good engineers are certainly more perceptive in economic matters than bad economists (and often, for that matter, good economists), anyone genuinely convinced that the world economy now enjoys a permanent surfeit of energy has seriously neglected his or her training in the "dismal" science. I must admit, however, that I belonged to the large group of experts mentioned in Gerondeau's book who did not believe that the present adjustment to a much lower intensity of energy use community could take place in such a short time. Contrary to my prediction, the citizens of the industrial world were prepared to accept a much higher level of unemployment and social insecurity than I believed possible only a decade ago. Making them accept these discomforts has been a kind of political miracle, although it might be also that many of the politicians responsible for this miracle are unaware of its ultimate costs. These costs include, for the United States at least, the partial dismemberment of the energy-intensive industries that were responsible for the industrial hegegemony of the United States during most of this century. In addition, there are the as-yet-unknown psychological and economic consequences that may result when average Americans discover that a high-tech heaven cannot be created in North America to restore the economic golden age that existed before the first oil price shock.

The other main European importers of coal are Italy, Spain, the Netherlands, Belgium, Denmark, and Finland. Although Italy provides a terminal for two huge natural gas pipelines from Algeria and the Soviet Union, the general opinion is that Italy will soon replace France at the top of the European coal-buying league. By the same token, Spain can be counted on to at least double its imports by the turn of the century; since Spain has an important domestic coal-producing capacity, there is no aversion to the use of coal in the electrical generation or industrial sectors. Furthermore, the present Spanish government is not friendly towards nuclear power, and there is a strong movement in favor of scrapping all plans to build further nuclear facilities.

Data Resources Incorporated has estimated that the Netherlands and West Germany will raise their coal imports by more than Spain; but consumption is rising at a much lower rate than predicted in Holland, and the use of natural gas has not been allowed to fall at the rate predicted five years ago.

As for West Germany, there is a strong desire on the part of the decision-makers to use more nuclear energy, even though the environmental movement—and many Germans not formally associated with this movement—oppose it. But with the environmental destruction now taking place in the German forests, it appears that "The Greens" are going to have to reformulate their strategy. This is not going to be easy, because if the environmentalists in Germany (and elsewhere) were more pro-environment, and less anti-industry, they would understand that the way to solve the environmental problem, and to maintain employment and economic growth, is to subsidize the installation of pollution-abatement equipment on coal burning installations, and to ignore some of the ivory tower prescriptions for reducing environmental damage that characterize much of the learned literature. A point of some interest is that it is less expensive to use South African coal bought in Rotterdam or Hamburg than German coal purchased at the mine mouth—at least most of the time. Accordingly, if further pit closures in Germany are to be avoided, which is ostensibly a part of the program of the Greens, then steps will have to be taken immediately to maintain the overall demand for coal. This, in turn, implies that desulfurization equipment will have to be added to many coal-burning power plants right away.

Coal Importers in the Pacific Region

On the other side of the world, the main importers are Japan and the newly industrializing countries of East Asia, particularly South Korea and Taiwan. South Korea buys mainly from Canada and Australia; the main purchasers are electricity-generating companies, especially Korea Electric. The main purchaser on Taiwan is Taipower, which has placed large contracts in Canada and South Africa. If it can be assumed that these two countries will continue to experience a high rate of economic growth, then their combined coal requirements by the year 1990 may reach 25 million tons.

As for Japan, that country is—and seems likely to remain—the world's top buyer of coal. Politically, at the present time, it seems possible for the Japanese to use an unlimited amount of nuclear energy; but, from time to time, there has been an extremely vigorous antinuclear movement in Japan, and another accident of the Three Mile Island variety in North America, Europe, or Japan would probably revitalize this movement. In these circumstances, the Japanese government has decided to make sure that if there were trouble on the nuclear front, alternative sources of energy would be available.

About one-half of Japan's coal comes from Australia. Another important supplier is the United States, and although the branch of microeconomics known as portfolio theory tells us why, it cannot give us any answers in so far as quantities are concerned. U.S. coal is the most expensive coal purchased by Japan, and clearly it would be simple to obtain the same amount of coal from Canada, Australia, and South Africa; however nobody knows what the world coal market would look like in the future, nor just how much coal Japan will need, and so in order to retain ties with U.S. sellers, Japan is willing to pay top prices for U.S. coal. Japanese buyers may also believe that if the Pacific market remains large enough, then the United States will make substantial investments in railroad and port faciltiies, which could lead to a decrease in the future price of U.S. coal. Well-informed sources have suggested that the Japanese are investing heavily in the west Canadian coal sector, including transportation facilities, for the purpose of keeping the world 'supply curve' for coal moving to the right. It has also been suggested that this could turn out to be a very attractive investment from the point of view of potential yield, particularly if labor relations in the Canadian coal mining sector do not worsen.

The Japanese Ministry of Industry and Trade recently estimated that Japan will be using about 110 million tons of coal in 1990, with 45.3 to 48.3 of these being steam coal. But this forecast represents a 30 percent decrease in their 1983 estimate. The Japanese have had extraordinary success in reducing coal-based emissions, and so pollution is less of a constraint on the use of coal in that country than in North America or Western Europe. However, the security problem associated with managing nuclear energy may be less complicated in Japan than elsewhere (assuming that nothing happens to disturb the present widespread acceptance of nuclear power), and so there is considerable pressure from decision-makers in government and industry to use as much nuclear energy as possible.

As an indication of its belief in diversification, Japan imports sizable amounts of coal from the Soviet Union and China, and has indicated a desire to obtain more. Also among their suppliers are Indonesia and New Zealand, each of which has provided Japan with miniscule amounts of both steam and coking coal during the past four years.

Two Exporters of Coal: Australia and South Africa

This book as a whole has taken particular care to present readers with an economic analysis of the world trade in coal, while in this chapter, one of the most important topics is the exporting of steam coal to Western Europe by the United States. This is because coal can only fulfill its destiny, as visualized by WOCOL and the IEA, if the United States shows itself capable of making huge amounts of coal available to the energy-deficient countries of Western Europe and Asia. But before examining this issue, a few things should be said about Australia and the Republic of South Africa.

Almost all Australian hard coal production (and all coal exports) comes from two states: Queensland and New South Wales; production may have reached a record level in each of these states in 1983. There has been, however, a change in the structure of coal production in Australia, in that the output of open cast (i.e., open pit) mines is increasing rapidly, while underground production is either stagnant or falling. This has tended to increase profits, but unions are upset because an increase in the output of less labor-intensive open cast mines, in a market that is not expanding, represents a threat to total mining employment. One of the most important comparative advantages enjoyed by Australian mines is that many of them are located near the ports from which coal is shipped to Asia or Europe. Port development has lagged somewhat in Australia, but this shortcoming is being remedied, and Australia may already be reaping the benefits of a well-thought-out program of transportation improvement. It should be appreciated, however, that a stagnant world coal market cannot be revived by investments in coal mines and shipping facilities. For a long time it had been expected that when Queensland's new export-oriented Blair Athol mine came on stream, it would radically alter the profits of Queensland's coal industry. Instead, after opening in the spring of 1984, Blair Athol has been involved in serious contractual conflicts with its customers. Japan's Electric Power Development Corporation and Japan Coal Development have refused to take the volume of coal previously contracted for, or pay the agreed price, and so coal is being shipped to Japan at a stop-gap price until further negotiations clarify the situation. The outcome of these negotiations is almost certain to mean an Australian retreat: the Japanese market is too valuable to Australian coal producers to risk its loss to Canadian and/or South African suppliers.

In 1974 South Africa exported hardly any steam coal. In a year or two from now, its annual exports may come to 40 million tons/year, or more. At present, South African producers are doing everything possible to make sure that they export the maximum amount of coal allowed by the South African government under its export license system. Current sales are regarded as disappointing by many mining companies; instead of the 38 or 39 million tons that South African coal sellers had hoped to export, they will have to settle for total exports of about 35 million tons. However, some observers claim that the coal market has turned up, and if this is true, it means that South Africa will soon find itself in a position to appreciably expand its export income.

The export of coal from South Africa is facilitated by the cooperation between South African coal producers and several of Europe's major energy companies: Royal Dutch Shell, British Petroleum, and Total (France). The largest coal producer in South Africa is Amcoal Ltd., a subsidiary of the Anglo American Corporation—the world's largest producer of gold. Apparently, many North American energy firms avoid business ventures in South Africa because they are afraid that political repercussions might damage their market prospects in the United States. This is also an important consideration in Europe, where companies like Shell make a point of not advertising the details of their involvement in South Africa.

One of the South African coal sector's great advantages is its export

port at Richards Bay, which is situated on the Indian Ocean north of Durban, about 300 miles from South Africa's main coal fields. This port is perhaps the best equipped, operated, and most modern in the world; Port Richards can probably handle more than 110,000 tons a day, but it is possible that by 1990 it will be too small for South Africa's export trade. Perhaps the lowest-cost coal in the world, on average, comes from South African open cast operations. The average cost of this coal at the mine mouth is said to be about $12 dollars per ton, which is increased to $25–$26 on board ship at Richards Bay. At the present time, this coal is being sold on European spot markets for $41 dollars per ton, as compared to $45 dollars per ton for Polish coal, which is the next cheapest. U.S. steam coal, at $57 dollars per ton, is no longer listed on the spot market.

The United States and the Soviet Union

Shortly after the first oil price shock, the International Energy Agency (IEA) and others advanced the opinion that, in the light of a growing scarcity of oil, the production of coal in the World Outside the Centrally Planned Areas (WOCA) should be at least doubled, and international trade in coal should expand accordingly. The basic reasoning forwarded by the IEA was that if the United States could realize its coal production and export potential, then governments, electricity-generating firms, and industrialists would see that a twenty- or thirty-year supply of reasonably priced coal was unconditionally available, and they would make the costly investments required to receive, transport, and burn this commodity.

A dynamic upgrading of export facilities began in the United States in the mid-1970s, and an export boom commenced in 1978 that enabled the United States to raise its exports of coal from 40 million tons of mostly coking coal to about 110 million tons of mostly steam coal between 1978 and January 1982. Revenues from the export of coal reached more than $6 billion, but the recession caused by the second oil price shock eventually terminated this upswing. Exports fell to 105 million tons in 1982, and then collapsed to just 77 million tons in 1983. Foreign sales will probably remain low during 1984, although production may rise to a record 850 million tons. Many coal industry officials do not expect this production level to be maintained in the immediate future, even though the National Coal Association expects the total demand for U.S. coal to rise to 1.1 billion tons a year by 1995, which is an increase of 28 percent in eleven years.

Before taking a closer look at some of the constraints on an explosive expansion of exports, a few optimistic remarks seem in order. After bottoming out in 1979, labor productivity in the U.S. coal industry—

which had fallen 32 percent during the previous decade—made a spectacular recovery in 1980. There was a temporary pause in the traditional bad relations between labor and management, and contract negotiations between employers and the United Mine Workers produced the first "peaceful" settlement since the early 1960s. Moreover, in the future, investments will almost certainly be made, sooner or later, in at least one deep-water port on the United States eastern seaboard, which will permit the loading of the largest colliers. This will mean an important cost reduction for coal being exported to Europe. In addition, shipping rates (and probably bunker fuel prices) will rise from their present extremely low levels, which will favor the United States relative to Australia, and perhaps South Africa.

Thus, the future prospects of the U.S. coal-exporting sector appear to turn on the ability of the market, or the United States Congress, to curb the avariciousness of the main coal-hauling railroads. At the present time, on two important eastern routes, coal being shipped to export terminals is charged a rail tariff of almost 16 dollars per ton, while coal traveling the same distance on barges—but intended for the domestic market—pays only 6 dollars per ton in transport costs. Lower rail rates would make all U.S. coal more competitive in foreign markets, but it would particularly favor highly salable low-cost, low-sulfur coal from the West and Midwest. Of greater importance, this coal could be obtained from larger deposits or mines than generally exploited in the East, and thus western firms could sign larger and longer-term contracts with buyers than, for example, eastern firms. This would mean that, unlike the shipments of producers selling via spot or short-term contracts, the shipments and revenues of these large mining firms would be less influenced over time by the prevailing price and, to a considerable extent, even by falling demand. Consequently, it might be possible for the railroads to more than make up in volume what they lose in price, which implies that they would not lose any revenue. Denmark's Elkraft has been attempting to obtain a special rail tariff of 10 dollars per ton for the coal purchased from eastern mines, instead of the 15 dollars per ton that is now being paid, but unfortunately the railways are not interested. As a result, Elkraft has no choice but to partially or totally cancel its contracts with A.T. Massey Coal Company and Island Creek Coal Incorporated, from whom it obtained 630,000 tons annually. Since Elkraft also buys coal from Poland, the Soviet Union, Australia, South Africa, and Britain at prices under that for U.S. coal, a canceling of their U.S. contracts and transferring this business to some or all of their other suppliers, would of course save them a great deal of money. The same thing is true for the government-controlled Italian steel firm Italsider, which wants to favor U.S. exporters, but is unable to do so because the price of U.S. coal makes it unacceptable in all except very small amounts. Naturally, cancellations mean losses for coal firms and railways.

This above dilemma is well known, and thoroughly discussed, in almost every publication that takes an interest in the U.S. energy sector. Almost equally important, but considerably less discussed, are the uncertainties associated with U.S. ability to produce, over a very long time horizon, a great deal of reasonably priced coal without a dramatic alteration in the present geographical pattern of production. The expansion in exports and production in the mid-1970s was accompanied by a significant increase in U.S. coal prices. This increase could hardly have been avoided, given the fact that in the small and medium-sized mines of the East and Midwest short-run costs often rise more than proportionally to output, and scale economies are virtually unobtainable at the present level of mechanization. In this context, the work of Professor Martin Zimmerman (1981), which deals with the effect of depletion on mining costs, is useful. With mainly low-sulfur coal entering into international trade, Zimmerman's analysis indicates that there will be a substantial rise in the cost of obtaining low-sulfur Appalachian steam coal not too much further along in its depletion cycle.

It also appears that the costs of midwestern and Appalachian medium-sulfur coal will increase, though not as much as for low-sulfur coals; but to some extent this is irrelevant for the export market, since Western European buyers are not likely to be interested in medium-sulfur coal in the near future. The news of the German forests has begun to attract a great deal of attention in Europe and, as pointed out in chapter 6 of this book, there seems to be decisive evidence that the damage to trees is accelerating—and most people think that coal is the reason. This leads immediately to the question of what can be done to remedy the situation without dispensing with coal. The Japanese have apparently been very successful in their war against atmospheric pollution, and much of the equipment and techniques they are employing can and will be used in other parts of the world. It has also been claimed that another very efficient technology for obtaining a large reduction in the discharge of sulfur dioxide is ASEA (of Sweden's) PFBC process. The important thing now is to invest in this technology, and I am personally convinced that this will not be a problem if the burden of financing is distributed equitably throughout the community.

I will close this summary with a few remarks about the Soviet coal industry. As pointed out in my book, *The Political Economy of Oil,* this sector does not always appear to have an especially bright future. The Five-Year Plan target for 1985 of 775 million tons is out of the question, and at present a great effort must be made to keep production fron stagnating.

Still, I have been informed that the directors of the Soviet coal industry have every intention of making the industry more effective and of eventually getting it to make an important contribution to Soviet export revenues. In order for this to take place, there will have to be a quantum increase in productivity, particularly in underground mining, while the pace at which new mines are being commissioned will have to be increased. This implies a sharp rise in investment in the coal sector, and there is some doubt as to whether this is possible. Clearly, sectors such as agriculture must be given top priority in the Soviet economic plan. There is also the matter of whether further capital should be allocated to coal production in the light of the requirements of the Soviet oil industry (whose 1.4–1.6 million barrels per day of exports to WOCA is the most reliable earner of hard currency in the Soviet Union) and perhaps also of the natural gas industry.

As in the rest of the world, underground mines in the Soviet Union have had a great deal of difficulty preventing rising costs. One of the main problems is that their mines tend to be much deeper than those of North America, for example. On the other hand open pit mines appear to be doing relatively well: better machinery is being provided, and apparently there is further scope for the reduction of costs if managers are more adept at adopting best-practice technologies. It may also be true that managers will have no choice about becoming more adept, since the Coal Ministry has been informed by the Soviet leadership that the career prospects of pit managers who fail to obtain the desired results should be limited.

The Soviet coal industry's center of gravity is slowly moving eastward, toward Siberia and Kazakhstan, and it has been suggested that this will result in a very large aggregate increase in Soviet coal production either late in this century, or early in the next. Whether this is true or not remains to be seen, because the problems associated with establishing an efficient coal mining industry in the frontier districts of the Soviet Union are, to put it mildly, tremendous. I also believe that the Soviets would like to receive a great deal of financial and technological help from Western Europe in order to facilitate this displacement and that it is likely they will receive it—though not in the immediate future.

Concluding Remarks

According to Lord Keynes, "The study of economics does not seem to require any specialized gifts of a very high order. Is it not, intellectually regarded, a very easy subject compared with the higher branches of philosophy and pure science? Yet good or even competent economists are the rarest of birds. An easy subject at which few excel." (From *The Collected Writings of John Maynard Keynes,* XIV, edited for the Royal Economic Society by Elizabeth Johnson. London: Macmillan and Cambridge University Press, 1972).

Regardless of who excels or fails to excel at the "dismal" science, one thing should not be forgotten: the problems economists have to solve today are many times more difficult than those addressed by that premier economist of this century. If our only concern were to provide formulae for full employment and expanding production against a background of inexhaustible energy and natural resources—and Keynesian economics—then we would have no trouble telling the decision-makers how to make this world a beautiful place. But, alas, such is not the case. Not only do we have to contend with oil price shocks, burgeoning pollution, and dubious concepts like monetarism and rational expectations, but just over the horizon is the most intractable impasse of all: the exploding world population. Economists with something intelligent to say about that dilemma are rare.

Thus, it is essential to train economists and politicians to comprehend that real life is subject to unexpected and ill-timed changes that are not usually at home in neo-classical theory. Arithmetic, hearsay, and daydreams cannot take the place of economic reasoning, but as things now stand, sensible people searching for solutions to economic problems should stop first on the plane of ethics.

Where coal is concerned, this observation leads to two interdependent recommendations. First, we have been warned at the highest government levels (e.g., the National Research Council of the United States) that environmental deterioration is carrying our planet into "largely unknown territory," which sounds to me like some kind of euphemism for a disaster. Taken by itself, the warning suggests that we will have to be extremely careful with the way that fossil fuels, and particularly coal, are burned. On the other hand, with unemployment in the OECD somewhere in the vicinity of 33 million people and poised for further increases, the industrial world is not in position to experiment with low-energy societies. Within the framework of stricter environmental controls, the consumption of coal can and must expand.

A Comment on the Literature and a Bibliography

The three most important academic journals for energy matters are *Energy Policy, The Energy Journal,* and *Resources Policy* (which, at the present time, is concentrating on nonfuel minerals). Anyone interested in staying up to date on energy matters should read *The Petroleum Economist,* The *Financial Times Energy Economist,* The *Quarterly Energy Review* (published by the Economist Intelligence Unit), and the *Oil and Gas Journal.* I also make a point of reading *The OPEC Quarterly Review, The OPEC bulletin,* and—for petrochemicals—the *Chemical Economy and Engineering Review* (Tokyo). Finally, the leading coal periodicals seem to be *Coal Age, Coal Week,* and *Coal Week International* (all published by McGraw-Hill), and *World Coal,* published by Miller Freeman Publications, San Francisco.

Abelson, P. *Coal in Australia.* Economist Intelligence Unit, Report No. 149, London 1983.

Andersson, R., and Bohm, P. *Samhällsekonomisk Utvärdering av Energieprojekt,* Stockholm: Nämnden för Energiproduktionsforsking, 1981.

Banks, F.E. "An Econometric Model of the World Tin Economy: A Comment," *Econometrica* 1972.

——— "A Note on some Theoretical Issues of Resource Depletion," *Journal of Economic Theory,* October 1974.

——— *The International Economy: A Modern Approach.* Lexington, Mass.: Lexington Books, 1979.

——— *The Political Economy of Oil.* Lexington, Mass.: Lexington Books, 1980.

——— Resources and Energy: An Economic Analysis. Lexington, Mass.: Lexington Books, 1983.

Baumol, W. J. *Economic Theory and Operations Research.* London: Prentice Hall International, 1977.

——— and Oates, W.E. *The Theory of Environmental Policy.* London: Prentice Hall International, 1975.

Bessière, F., and Molat, G. *Vingt Cinq Ans d'Economie Électrique. Paris: Dunod, 1971.*

Bohm, P. "CFC Emissions in an International Perspective," in *The Economics of Managing Chlorofluorocarbons,* edited by J. H. Cumberland and others. Washington: Resources for the Future, 1982.

Boiteaux, M. "La Tarification de Demandes en Pointe." *Revue Générale de l'Electricité,* No. 58, 1949.

——— "La Tarification au Cout Marginal et les Demandes Aléatoires." *Cahiers du Seminaire d'Econométrie,* 1951.

Carling, A., Björk, O., and Kjellman, S. *Internationella Energi Marknader: Prognos och Framtids Bedömningar.* Stockholm: Industrie Utrednings Institut, 1979.

Davies, T. A. "The Outlook for Natural Gas in Western Europe to 2000," stencil, 1984.

Desprairies, P. *La Crise de l'Energie.* Paris: Editions Technip, 1982.

Devins, D. W. *Energy: Its Physical Impact on the Environment.* New York: John Wiley and Sons, 1982.

Edmonds, J., and Reilly, J. "A Long-Term Global Energy-Economic Model of Carbon Dioxide Release from Fossil Fuel Use," *Energy Economics,* April 1983a.

——— "Global Energy Production and Use to the Year 2050," *Energy,* June 1983b.

——— "Global Energy and CO_2 to the Year 2050," *The Energy Journal,* October 1983.

Einhorn, M. "Optimal System Planning with Fuel Shortages and Emissions Constraints," *The Energy Journal,* April 1983.

Estrada, J. "The Importance of the Natural Gas Industry in the Soviet Union," stencil, Bedriftsøkonomisk Institutt, Oslo, 1984.

Frazer, F. *Gas Prospects in Western Europe.* London: Financial Times Business Information, 1982.

Fettweiss, G. B. *World Coal Resources: Methods of Assessment and Results.* Amsterdam: Elsevier-North Holland, 1979.

Folie, M., and McColl, G. "An Economic Appraisal of the Australian Coal Industry," stencil, Macquarie University, Sydney, 1978.

Fritsch, B. "Über die Partielle Substitution von Energie, Resourcen, und Wissen," Stencil, Mannheim, 1979

Fugelberg, O. "Demand for Natural Gas in Western Europe: The Case of France," Stencil, Bedriftsøkonomisk Institutt, Oslo, 1984.

Gerondeau, C. L'Energie à Revendre. Paris: Lattès, 1984.

Gihel, J. "La Tunique de Nessus." *Revue de L'Energie,* November 1981.

Giraud, A. "Les Raisons d'une Déroute," *Le Figaro,* 5 February, 1982.

Gollop, F. M., and Roberts, M. J. "Environmental Regulations and Productivity Growth: The Case of Fossil-Fueled Electric Power Generation," *Journal of Political Economy,* No. 4, 1983.

——— "Efficient Regulation of Sulphur Emissions: Regional Gains in Electric Power," stencil, July 1983.

Gordon, R. *Coal in the United States Energy Market.* Lexington, Mass.: Lexington Books, 1978.

Granström, T. *Elförsörjningen i Energiedebattens Centrum.* Stockholm: Liber, 1982.

Grathwohl, M. *World Energy Supply: Resources, Technologies, Perspectives.* New York: Walter de Gruyter, 1982.

Hansson, B. "Svavlet Kan Elimineras," *Svenska Dagbladet,* 24 Juli, 1981.

Harlinger, H. "Neue Modelle Für die Zukunft der Menschheit," IFO Institut für Wirtschaftsforschung, Munich, 1975.

Hellman, R., and Hellman, C. J. C. *The Competitive Economics of Nuclear and Coal Power.* Lexington, Mass.: Lexington Books, 1983.

Hoffmeyer, M., and Neu, A. "Zu den Entwicklungsaussichten der Energie Märkte," *Die Wirtschaft,* Heft 1, 1979.

Houdaille, M. "Le Marché de l'Uranium," *Bull. Information Nucléaire,* July-August 1970.

James, P. *The Future of Coal.* London: Macmillan, 1982

Janssen, E. R. "Le Prix du Petrole Brut et Les Produits Pétroliers et Leur Evolution en Europe," *Recherches Economiques de Louvain,* March 1978.

Joskow, P. L., and Baughman, M. L. "The Future of the United States Nuclear Energy Industry," *The Bell Journal of Economics,* Spring 1976.

Kjellman, S. "Handel Med Ångkol." Stockholm: Forskningsgruppen för Energisystemstudier, 1981.

Knudsen, K. "Den Glemte Årsak til Oljeprisens Mulige Kollaps." *Sosial Økonomen,* 1984.

Laffont, J. J. *Cours de Theorie Microeconomique* (Vol. 1: Fondements de l'Economie Publique), Paris: Economica, 1982.

Lave, L. "Factors Affecting Cooperation in the Prisoners Dilemma," *Behavioral Science,* 1965.

Long, R. *Constraints on International Trade in Coal.* IEA Coal Research, London, 1982.

Lütkenhorst, W., and Minte, H. "Probleme des Monetären und Realen Transfers," in M. Tietzel, ed., *Die energiekrise: Funf Jahre Danach,* Bonn: Neue Gesellschaft GmbH. 1978.

Malinvaud, E. *Lecons de Théorie Microéconomique.* Paris: Dunod, 1969.

Manabe, S. "Estimates of Future Change of Climate Due to an Increase in CO_2 Concentrations in the Air," in W. Mathews, W. Kellogs, and G. Robinson (eds.), *Man's Impact on Climate.* Cambridge: MIT Press, 1971.

——— and Wetherald, R. T. "The Effect of Doubling CO_2 Concentration on the Climate of a Generation Circulation Model," *Journal of Atmospheric Science,* No. 32, 1975.

Martellaro, J. A. "Soviet Energy Resources: Present and Perspective," *Economia Internazionale,* Spring 1981.

Munasinghe, M., and Warford, J. J. *Electricity Pricing.* Baltimore: The Johns Hopkins University Press, 1982.

Mårtensson, M. "Uran för Kraftalstring," *Teknisk Tidskrift,* 1967.

Neu, A. D. "Entkoppeling von Wirtschaftswachstum und Energiverbrauch: Eine Strategie der Energiepolitik?" Kieler Diskussions Beitrage, 1981.

Nordhaus, W. D. "The Allocation of Energy Resources," *Brookings Papers,* No. 3, 1973.

——— "How Fast Should We Graze the Global Commons," *American Economic Review,* May 1982.

Oboussier, F. "Die Versorgung der Bundesrepublik Deutschland mit Angerichertem Uran," *Glückauf,* 1976.

Olby, R. "Britain's Resources of Coal and Spent Uranium Fuel," *Nature,* 24 April, 1982.

Owen, A. D. "The Economics of Uranium Demand," *Resources Policy,* June 1983.

Peacock, C. G. "The Political Economics of the Gas Supply Options Open to Western Europe," stencil, Chemical Systems International, London.

Percebois, J. "L'Europe et la Cooperation Internationale dans le Domaiune des Hydrocarbures Gazeux," stencil, Institut Economique et Juridique de l'Energie, 1983.

Prast, W. G., and Lax, H. L. *Oil-Futures Markets.* Lexington, Mass.: Lexington Books, 1983.

Prior, M. *Steam Coal and Energy Needs in Western Europe Beyond 1985.* Economist Intelligence Unit Special Report No. 134, London, 1982.

Radetzki, M. *Uranium: A Strategic Source of Energy.* London: Croom-Helm, 1981.

Robinson, C., and Morgan, J. *North Sea Oil in the Future.* London: Macmillan, 1979.

Romasset, J., Isaak, D., and Fesharaki, F. "Oil Prices without OPEC: A Walk on the Supply Side," *Energy Economics,* July 1983.

Sargent, D. A. "The United States Role in the International Thermal Coal Market," *The Energy Journal,* January 1983.

Sauter-Servaes, F. "Der Übergang von einer Erschöpfbaren Ressource zu einem Synthetischen Substitut," Mannheim: Vereins für Socialpolitik, Gesellschaft für Wirtschafts-und-Sozialwissenschaften, 1979.

Schultz, W. "Die Langfristige Kostenentwicklung für Steinkohle am Weltmarkt," *Zeitschrift für Energiewirtschaft,* No. 1, 1984.

Solow, R. M. "The Economics of Resources or the Resources of Economics," *American Economic Review,* May 1974.

Spengler, J. "Population and World Hunger," *Rivista Internazionale de Scienze Economiche E. Commericali,* December 1976.

Späth, F. "Die Preisbildung für Erdgas," *Zeitschrift für Energiewirtschaft,* No. 3, 1983.

Svenska Statens Offentliga Utredningar: Pris På Energi (med bl.a. O. Kun. Hjalmarsson) Industridepartementet, 1981.

Tempest, P. *International Energy Markets.* Proceedings of the 1982 Cambridge Conference of the IAEE, Cambridge, Mass.: Oelgeschlager, Gunn & Hain, 1983.

Wenders, J. T. "Peak Load Pricing in the Electric Utility Industry," *Bell Journal of Economics,* Spring 1976.

Wilson, C. *Coal: Bridge to the Future*. Cambridge, Mass.: Ballinger Publishing Co., 1980.

Woodwell, G. M. "The Carbon Dioxide Question," *The Scientific American*, January 1978.

Zimmerman, M. B. *The United States Coal Industry: The Economics of Policy Choice*. Cambridge, Mass.: MIT Press, 1981.

Index

About the Author

Ferdinand E. Banks is Social Sciences Faculty Research Fellow at the University of Uppsala, and visiting professor at the International Graduate School, the University of Stockholm. In 1983 he was visiting research fellow at the Institute for Applied Social and Economic Research, the University of Melbourne; and visiting lecturer at the Australian School of the Environment, Griffith University, Brisbane Australia. In 1981 he held a Tore Browaldh Foundation Fellowship for travel and research in Australia and Germany; and during 1980 he was visiting professor at the Centre of Policy Studies, Monash University, Melbourne, Australia. In 1978–1979 he was professorial fellow in economic policy at the Reserve Bank of Australia, Sydney Australia, and visiting professor in the Department of Econometrics, the University of New South Wales.

Professor Banks attended Illinois Institute of Technology and Roosevelt University, Chicago, Illinois, and received a B.A. in economics. After serving with the United States Army in the Orient and Europe, he worked as a systems and procedures analyst, and as a civilian employee for the United States Navy. He received M.Sc. and Fil. Lic. degrees from the University of Stockholm, and the Fil. Dr. from the University of Uppsala, where he has taught since 1971.

Professor Banks has also been a lecturer at the University of Stockholm for five years; a senior lecturer in mathematics, economics, and statistics at the United Nations African Institute for Economic and Development Planning, Dakar, Senegal; and has been consultant lecturer in macroeconomics and input–output analysis for the OECD at the Technical University of Lisbon, Portugal. From 1968 to 1971 he was an econometrician and primary commodity economist for the United Nations Commission on Trade and Development (UNCTAD) in Geneva, Switzerland. He has also been consultant for the United Nations Development Program in Cairo and Addis Ababba, the Hudson Institute (Europe) in Paris, France, and consultant on planning models and the steel industry for the United Nations Industrial Development Organization (UNIDO) in Vienna. His previous books are *The World Copper Market* (1974), *The Economics of Natural Resources* (1976), *Scarcity, Energy, and Economic Progress* (1977), *The International Economy: A Modern Approach* (1979), *Bauxite and Aluminum: An Introduction to the Economics of Non-Fuel Minerals* (1979), *The Political Economy of Oil*

(1980), and *Resources and Energy* (1983). He has also published ninety-two articles, notes, reviews, and other communications in various journals and collections of essays. He is currently working on the economics of refining and petrochemicals, international monetary problems, the futures market for oil and oil products, and the economics of the environment.